T0399381

A Thermo-Economic Approach to Energy From Waste

A Thermo-Economic Approach to Energy From Waste

ANAND RAMANATHAN
Department of Mechanical Engineering, National Institute of Technology, Tiruchirappalli, India

MEERA SHERIFFA BEGUM K.M.
Department of Chemical Engineering, National Institute of Technology, Tiruchirappalli, India

AMARO OLIMPIO PEREIRA JUNIOR
Department of Energy Planning - PPE/COPPE, Federal University of Rio de Janeiro - UFRJ, Rio de Janeiro, Brazil

CLAUDE COHEN
Faculty of Economics and Administrative Justice Post-graduation Program - PPGJA, Federal Fluminense University - UFF, Rio de Janeiro, Brazil

ELSEVIER

Elsevier
Radarweg 29, PO Box 211, 1000 AE Amsterdam, Netherlands
The Boulevard, Langford Lane, Kidlington, Oxford OX5 1GB, United Kingdom
50 Hampshire Street, 5th Floor, Cambridge, MA 02139, United States

British Library Cataloguing-in-Publication Data
A catalogue record for this book is available from the British Library

Library of Congress Cataloging-in-Publication Data
A catalog record for this book is available from the Library of Congress

ISBN: 978-0-12-824357-2

For Information on all Elsevier publications
visit our website at https://www.elsevier.com/books-and-journals

Publisher: Candice Janco
Acquisitions Editor: Peter Adamson
Editorial Project Manager: Chris Hockaday
Production Project Manager: Debasish Ghosh
Cover Designer: Mark Rogers

Typeset by MPS Limited, Chennai, India

Contents

About the authors

Dr. Anand Ramanathan is an associate professor in the Department of Mechanical Engineering at the National Institute of Technology, Trichy. Area of specialization involves internal combustion engines, alternative fuels, waste-to-energy conversion, emission control, and fuel cells. He is the recipient of the Australian Endeavour Fellow and worked on solar fuels at Australian National University at Canberra, Australia, from July to October 2015. He has received various sponsored projects from GTRE-DRDO, DST-SERB, DST-YSS, DST-UKERI, MHRD, and BRICS. He has received Indo-Brazil collaborated project in the area of the thermochemical conversion process with life cycle assessment under the SPARC scheme. He has contributed several paper publications in renowned international journals. He has filed seven Indian patents. He has also contributed 12 book chapters in a renowned publication (Elsevier and Springer), and involved in professionally related activities and administrative responsibilities to serve the community.

Dr. Meera Sheriffa Begum K.M. graduated from Anna University, Chennai. She has got industrial experience at "Chennai Petroleum Corporation Ltd" (formerly Madras Refineries Ltd) at R&D division as "MRL Research Fellow" from 1991 to 1995. She is currently a professor at the Department of Chemical Engineering. She has received many best paper awards in international conferences, granted two patents, published several publications in renowned international journals, and coauthored three chemical engineering textbooks published by PHI, India. She has contributed research, sponsored, and consultancy projects toward sustainable environment and energy funded by MHRD, DST and CSIR, and SPARC, and undertaken research training at NUS, Singapore, through TEQIP. She has executed industrial consultancy projects for wastewater treatment in Trichy Distilleries, dairy industry, textile industry, and Tamil Nadu Paper, Ltd, using functional materials toward sustainability and, also, blended biodiesel development and its execution on Southern Railways.

Prof. Amaro Olimpio Pereira Junior is an economist, PhD in energy planning from the Federal University of Rio de Janeiro. He worked as a technical advisor of the Energy and Environment Department at Energy Research Company (EPE), in Brazil, is a visiting professor at University Pierre Mendès-France in Grenoble, France, and at the University of Texas at Austin, Texas. He also worked as Research Fellow at CIRED (Centre International de Recherche sur l'Environement et Dévélopement), France. Currently, he is an associated professor of the Energy Planning Program of the Institute of Graduate Studies in Engineering at Federal University of Rio de Janeiro (PPE/COPPE/UFRJ), a researcher at CentroClima, and a member of the permanent technical committee at LIFE. He is an author of books, book chapters, and several papers in international journals.

Prof. Claude Cohen obtained her PhD in energy planning from the Energy Planning Program at The Federal University of Rio de Janeiro - PPE/UFRJ and a Master's degree in Development Economics from the University of Paris X Nanterre. She is currently an associate professor at the Faculty of Economics at Federal Fluminense University - UFF, a permanent professor at the Administrative Justice Postgraduate Program of Federal Fluminense University - PPGJA/UFF, and the coordinator of the Innovation, Environment and Sustainability Research Center of Federal Fluminense University - NIMAS/UFF. She has experience in economics, with emphasis on environmental and energy economics, working mainly on the following themes: sustainable development, climate change, energy consumption patterns, energy consumption in favelas, entrepreneurship and sustainable innovation. She is the author of book chapters, and several papers in international journals.

Preface

The authors of this book have been encouraged in their writing by long-term systematic attempts to connect research and development achievements with real-world applications. The waste-to-energy strategy is a hot topic that deserves to be explored. People are committed to create various sorts of wastes until the world's economic condition improves to a reasonable level. In this regard, the conversion of waste resources into value-added products combined with a lifecycle assessment to evaluate the actual environmental impact on thermochemical conversion processes.

The authors would like to thank Scheme for Promotion of Academic and Research Collaboration, SPARC-MHRD (Project No SPARC/2018-2019/P965/SL) for providing valuable resources and support for the successful completion of this book.

Many industry persons and academic experts supported us in reviewing draft chapters throughout the production of this new edition and created new graphics. We would like to specifically acknowledge the following peoples: Mr. M. Dineshkumar, Mr. R. Muthu Dinesh Kumar, Mr. G. Vigneshwaran, Mr. Guilherme Rodrigues Lima, and Mrs. Giovanna Ferrazzo Naspolini who has contributed novel ideas, useful data, and statistics for preparing this book.

The inspiration for this book is to cover waste-to-energy conversion techniques from the initial concept to the final product based on recent results of long-term planning. The life cycle assessment technique aids in the design and development of an eco-friendly process that minimizes greenhouse gas emissions. Waste processing has the potential to become a source of energy that may be used to generate power, cooking, and transport fuels or in fuel cell applications.

We believe that this book will give enough information on the action that must be followed to implement the waste-to-energy project in the future.

Anand Ramanathan[1], Meera Sheriffa Begum K.M.[2], Amaro Olimpio Pereira Junior[3] and Claude Cohen[4]

[1]Department of Mechanical Engineering, National Institute of Technology, Tiruchirappalli, India

[2]Department of Chemical Engineering, National Institute of Technology, Tiruchirappalli, India

[3]Department of Energy Planning - PPE/COPPE, Federal University of Rio de Janeiro - UFRJ, Rio de Janeiro, Brazil

[4]Faculty of Economics and Administrative Justice Post-graduation Program - PPGJA, Federal Fluminense University - UFF, Rio de Janeiro, Brazil

Acronyms and abbreviations

TCC	Thermochemical conversion
TRACI	Tool for the Reduction and Assessment of Chemical and Other Environmental Impacts
$CH_3CH_2CH_2CH_2COOH$	Valeric acid
H_2SO_4	Sulfuric acid
S	Sulfur
H_2O	Steam
SRT	Solids retention time
NaOH	Sodium hydroxide
Na_2CO_3	Sodium carbonate
RT	Retention time
RKS	Redlich–Kwong–Soave
CH_3CH_2COOH	Propionic acid
lb	Pound
PENG–ROB	Peng–Robinson cubic equation of state
O_3	Ozone
O_2	Oxygen
NO_x	Oxide of nitrogen
OLR	Organic loading rate
NO_2	Nitrogen dioxide
N	Nitrogen
HNO_3	Nitric acid
MFV	Minimum fluidization velocity
Mt	Million tons
mm	Millimeter
mL	Milliliter
CH_4	Methane
MJ/m^3	Megajoule per meter cube
MT	Mega tonnes
LHV	Lower heating value
LCM	Lignocellulosic material
LCA	Life cycle assessment
kg	Kilogram
IGCC	Integrated gasification combined cycle
CML	Institute of Environmental Sciences
H_2	Hydrogen
HCl	Hydrochloric acid
HC	Hydrocarbon
HRT	Hydraulic retention time
GHGs	Greenhouse gases
g/m^3	Gram per meter cube

GTI	Gas Technology Institute
FERCO	Future Energy Resources Corporation
FBG	Fluidized-bed gasifier
FB	Fluidized bed
FBG	Fixed-bed gasifier
FT	Fischer–Tropsch
ft.	Feet
EB	Enriched bed
DB	Double bed gasification
DME	Dimethyl ether
CFB	Circulating fluidized bed
Cl	Chlorine
CO	Carbon monoxide
CO_2	Carbon dioxide
C	Carbon
$CH_3CH_2CH_2COOH$	Butyric acid
BFB	Bubbling fluidized bed
$Btu/ft.^3$	British thermal unit per cubic foot
BCC	Biochemical conversion
BCL	Battelle Columbus Laboratory
NH_4	Ammonium
NH_3	Ammonia
CH_3COOH	Acetic acid
BTL	Biomass to liquid
BG–FT	Biomass gasification-Fischer tropsch
CHP	Combined heat and power
IGCC	Integrated gasification combined cycle
NPV	Net present value
1G	First generation
EC	Energy cane
CS	Conventional sugarcane
°C	Degree centigrade
FTIR	Fourier-transform infrared spectroscopy
cm^{-1}	Per centimeter
CH_2	Methylene
NMR	Nuclear magnetic resonance
ppm	Parts per million
C6	C6 alditol
β-O-4	Lignin structure
wb	Wet basis
mg/Nm^3	Milligram per nanometer cube
SO_2	Sulfur dioxide
IC	Internal combustion
m/s	Meter per second
H_2O	Water
kJ/g	Kilojoule per gram

MJ/kg	Megajoule per kilogram
kJ/g mol	Kilojoule per gram per mole
kJ/mol	Kilojoule per mole
HV	Heating value
FTS	Fischer–Tropsch synthesis
Fe	Iron
Cu	Copper
FeO	Iron oxide
D2	Isotopes of hydrogen
T	Temperature
P	Pressure
SBCR	Slurry bubble column reactor
bbl	Barrel
IKIE	Inverse kinetic isotope effect
kW	Kilowatt
MJ/Nm^3	Megajoule per nanometer cube
$CaCO_3$	Calcium carbonate
$MgCO_3$	Magnesium carbonate
HCN	Hydrogen cyanide
H_2S	Hydrogen sulfide
$E(j)$	Bioenergy potential of *n* crops at the *j*th state
$CRs(i,j)$	Surplus residue potential of the *i*th crop at the *j*th state
$	Dollar
C_nH_{2n+2}	Alkane
C_nH_{2n}	Alkene
WGS	Water gas shift reaction
A	Food in units
AGSN	Subnormal clusters
AP1	Central Planning Area 1
CE/EU	European Union
D	Total number of households surveyed
E	Packaging in units
IRT	Total volume of irregularly disposed waste
NGOs	Nongovernmental organizations
Pc	Measured weight for food residues in gram
Pe	Package specific weight in gram
R1	Total households disposing their waste in the waste basket
R2	Total households disposing their waste in the dumpster
UPP	Pacifying Police Unit
VRI	Percentage of households that answered yes to incorrect waste disposal

CHAPTER 1

Pyrolysis of waste biomass: toward sustainable development

1.1 Introduction

In the next 20 years extensive consumption of petroleum resources will rise at a constant rate of 1.6% per year [1]. Since their depletion rate is higher than their regeneration rate, petroleum resources are regarded as valuable natural resources. Excessive use of petroleum products results in harmful pollutants, such as nitrogen oxides, sulfur dioxide, and carbon dioxide emissions, which are causing a major environmental issue [2]. Due to the burning of fossil fuels, greenhouse gases, such as carbon dioxide, are emitted into the atmosphere causing global warming [3]. According to the survey, it is reported that petroleum-based resources worldwide will get exhausted after 2042, which does not depend on the growth of petroleum oil consumption [4]. The limitations of petroleum resources are primarily to blame for these issues. Many new projects have recently begun to find new eco-friendly energy resources for future generations to reduce emissions and minimize the energy crisis [3]. Renewable energy supplies not only eliminate harmful environmental effects but also eliminate reliance on petroleum resources [5]. In fact, the recent requirement for renewable fuels is produced from starch-based resources, competing with other edible feedstocks [6]. Ideally, biofuels obtained from lignocellulose biomass, such as grass, woods, agricultural residue, and energy crops, are preferred to compete with already existing energy sources [7]. The lignocellulosic biomass avail abundantly, and it is considered a renewable resource. Globally, it is estimated that lignocellulosic biomass of 220 billion tons produced [4]. As compared to petroleum resources, lignocellulosic biomass is considered a carbon-neutral resource and also helps to mitigate global warming [8]. In economic view, lignocellulose biomass feedstock is more inexpensive in contrast with edible biomass, such as corn starch [9]. Hence, appropriate technology implementation will help to produce large and better quality biofuels [3]. Biochemical conversion technologies using lignocellulose as raw material are not economical and cost-effective. Nevertheless, thermochemical conversion techniques are cost economical [10]. Pyrolysis is the most feasible and cost-effective way to generate liquid fuel from biomass of all thermochemical conversion technologies [11]. In pyrolysis, degradation of biomass in an inert atmosphere at an elevated temperature of 400°C–600°C occurs.

A Thermo-Economic Approach to Energy From Waste
DOI: https://doi.org/10.1016/B978-0-12-824357-2.00005-X

The main product of pyrolysis is biomass, also known as bio-oil, which can be obtained by condensing hot pyrolysis vapors. During the pyrolysis of biomass, due to the large number of primary and secondary reactions, a condensable mixture of chemicals is formed as pyrolysis oil. The pyrolysis oil is complex mixture of about 300 oxygenated compounds [12,13]. Pyrolysis oil mainly comprises three important groups of compounds, such as lignin-derived compounds, carbonyl compounds, and sugar-derived compounds. Higher water content (around 15–30 wt.%) present in the pyrolysis oil results in lower calorific values [14,15]. The large quantity of carboxylic acid makes the pyrolysis oil with lower pH values around 2–2.5 [16]. The instability of pyrolysis oil is associated with more oxygen in nature; it provides primary variation between pyrolysis oil and hydrocarbon (HC) fuels [14,17]. Because of these unfavorable properties, such as higher reactivity, higher acidity, higher viscosity, and lower heating values [14]. In today's engine system, use of bio-oil as transport fuel is not possible. It is important to increase the quality of pyrolysis oil to make it comparable to HC fuel [18]. Efforts to remove the higher oxygen content in pyrolysis oil using some important upgradation technologies are required [14]. Different investigations have been conducted to acquire this objective through various upgradation technologies. Among other techniques, hydrodeoxygenation and catalytic cracking are widely explored as upgradation techniques [19]. In the pyrolysis process, the inclusion of a solid acid catalyst without the use of hydrogen in an ambient pressure environment is known as catalytic cracking [20]. Hydrodeoxygenation of pyrolysis oil produces desired HCs in a pressurized environment using metal catalysts [21].

1.2 Component of lignocellulosic biomasses

To understand the mechanisms of catalytic copyrolysis, it is required to have some basic knowledge about the characteristics of lignocellulosic biomass. Therefore this section will be dedicated to a summary of the properties of the same. It is known from past research that lignocellulosic biomasses are the types of complex biopolymer. The major constituents include complex cellulose, lignin, and hemicellulose.

Normally, by weight, the biomass contains approximately 45% of cellulose, 25% of hemicellulose, and 25% of lignin [6]. Cellulose primarily acts as a structural frame for the lignocellulose cell walls. By nature, cellulose is a straight chained saccharide polymer of glucose with a strong β-1,4-glycoside bond [22]. Owing to the inclusion of a number of hydroxyl groups within the polymer chains, cellulose also demonstrates the existence of several hydrogen bonds [23]. The fundamental structure of cellulose is considered to be made up of several amorphous and crystalline areas [24]. In addition to this, many fibers like strands of cellulose are further linked by hemicellulose and/or pectin and have a coating of lignin [23]. When compared to cellulose, hemicellulose has largely different properties in that it is more amorphous in structure and also has a random heterogeneous type composition. Hemicellulose polymers are constituted by multiple pentoses (such as arabinose and xylose), hexoses (fucose, rhamnose, galactose, and glucose among others), and uronic acids (such as methyl glucuronic acid, galacturonic acid, and glucuronic acid) [6]. The formation of a system of cellulose microfibrils and lignin is promoted by the presence of many short, branched hemicellulose chains, which in turn causes the resulting lignocellulose matrix to be extremely stiff [23].

In terms of physical properties, lignin was found to be insoluble in water and has neutral optical properties. It is also the second abundant organic composition in nature. Complex aromatic and hydrophobic amorphous biopolymers of propyl-phenol groups are the main constituents [7]. Coniferyl, *p*-coumaryl, and sinapyl alcohols are the three basic phenol-containing components. It is made of three basic phenol-containing components: sinapyl alcohols, *p*-coumaryl, and coniferyl [25]. These units are linked together by C—C (β-5, 5—5, β-1, and β—β linkages) and C—O (β-O-4, α-O-4, and 4-O-5 linkages) [26]. Another important function of lignin is the linking between cellulose and hemicellulose. This is achieved to give the cell walls a rigid and inflexible three-dimensional structure. Lignin was also found to store close to 40% of the total energy of lignocellulosic biomasses. This may be attributed to a high carbon content [25]. Normally, softwood usually contains a higher fraction of lignin as compared to hardwood and other agricultural residues [23].

The most abundant nonedible biomass is lignocellulosic biomass. According to estimates, sustainable lignocellulosic biomass of 220 billion dry tons is generated annually around the world. In America alone, lignocellulosic biomasses of 1.3 billion dry tons are produced annually, which is 50% of the carbon consumption in the form of gasoline and diesel may be substituted. In contrast to petroleum fuel sources, lignocellulosic biomass is often viewed as a carbon-neutral source. This means that it is extremely useful in the reduction of the effects of global warmings. In terms of the financial perspective, making use of lignocellulosic biomass as a starting point to manufacture biofuels is much cheaper than using corresponding edible biomasses and petroleum sources. As a result, to have an efficient production of large quantities of biofuels from lignocellulosic biomasses, it is necessary to make use of the appropriate techniques and technologies.

1.2.1 Cellulose

Cellulose plays an important part in the structural strength of the cell walls in green plants and algae. It is also known to be among the most plentiful polymers of organic nature. The basic structure can be defined as a straight chain saccharide of repeating pyranose units, which are joined by means of acetyl linkage (β,1—4-glycoside bonds). Owing to its lengthy straight chained structure, cellulose is not easily soluble in water.

Multiple chains of cellulose polymers are further linked to form strand like fibers. These are interlinked by hydrogen bonds to form microfibrils of cellulose. These extensive microfibrils are meshed in an intertwined manner, which is the reason for the rigidity of cell walls. Cellulose fibers are coated with hemicellulose and lignin fibers. The interconnection of the three hydroxyl groups in the pyranose rings causes hydrogen bonds to form both within the molecule and between two neighboring molecules. The presence of these extensive hydrogen bonds is one of the primary reasons for the crystalline shape of cellulose, which promotes its high strength and chemical inertness [27–29].

1.2.2 Hemicellulose

Hemicelluloses are heterogenous saccharide chains, constituted by various monomers, such as arabinose and galactose. Some other constituents include xylose, glucose, and mannose. These monomers link together to form many polysaccharide structures, such as glucuronoxylan and xyloglucan. In Angiospermic plants, the basic constituent of hemicellulose is usually xylan, while glucomannan constitutes hemicellulose in gymnosperms [30]. Hemicellulose has a lesser extent of polymer formation as compared to cellulose, with each molecule of hemicellulose containing anywhere between 50 and 200 monomers. The abundance in nature hemicelluloses fall behind cellulose, constituting up to 30% of organic matter in plants. Hemicelluloses also exhibit amorphous characteristics in contrast to crystalline celluloses and, as a result, are easily soluble in dilute acids and bases by hydrolysis [31].

1.2.3 Lignin

Research has revealed that lignin may be up to one-third of the organic constituents of wood and other tracheophyte plants [32]. In terms of structure, lignin has an extensively branched 3-D structure of phenyl groups lined to propane, with the majority of bonds being the aryl-alkyl ether bonds. Based on the amount of methoxyl groups in the constituent unit cells of phenylpropane, they can be subdivided into syringyl, guaiacyl, and *p*-hydroxyphenyl units. While lignin contains a variety of linkages, such as ester, carbon-to-carbon, and ether, the predominantly occurring bonds, as discussed above, are the ether bonds between an aromatic ring and a phenylpropane side chain (α-O-4, β-O-4, γ-O-4). These bonds also occur between two benzene rings, or also between two phenylpropane chains (α-O-β0, α-O-γ0). Carbon-to-carbon bonds are the next most concentrated linkages, with a variety of linkage types, such as 5–5, β-5, or α-β0 among others. The ester bonds, which are relatively few in number and found mostly in soft and herbaceous plants, make up the majority of the remaining bonds [32]. Similar to hemicellulose, lignin also tends to display a strong repulsion toward water, partly due to its aromatic nature. Lignin usually occurs in the gaps between celluloses and hemicelluloses and accounts for the strength and inflexibility of biomass [31]. Further studies show that lignin content also varies by the type of plant and the climates they grow. Lignin constitutes about one-third of the organic content in softwoods. This number reduces to about 25% in temperate hardwoods and 30% in tropical-zone hardwoods; the compression wood contains the highest lignin content with up to 40%, whereas in tension reaction woods in angiosperms, lignin content drops to at most 20% [33].

1.2.4 Ash

Minerals and other inorganic compounds, such as sulfur, sodium, silicon, potassium, phosphorus, magnesium, chlorine, calcium, and others, can be found in any biomass. In addition to this, biomasses may contain traces of heavier metals, such as cobalt and nickel [34]. It is very difficult to quantify the amount of inorganic content for every types of biomass due to the small content and wide variations; however, the general trend that is observed is the wastes from farming activities have a higher mineral content than biomasses obtained from wood sources [35].

1.2.5 Extractives

In addition to the above-mentioned constituents, biomasses tend to contain a large quantity of extractives. These are components that may be readily removed using non-reacting solvent (water, alcohol, etc.) and are mainly formed as a by-product of the metabolism processes of a plant. Primary metabolic extractives are usually simple compounds, such as organic acids, fats, and sugars. In contrast, secondary metabolic by-products are much more complex in nature. This complexity lends the advantage of being used in taxonomical nomenclature, as the by-products are unique to the enzyme reactions that produce them. Some commonly occurring secondary extractives are coumarins, monoterpenes, and acetogens [36]. Even though extractives usually constitute less than one-tenth of the total biomass, they play an integral role in increasing the heating value of biomass with their highly volatile nature. As a result, there is no major thermal difference between the origin biomass and the final residues when phenolic compounds are formed by lignin decomposition [37–39].

1.3 Types of pyrolysis

Biomass pyrolysis was categorized as slow, medium, or fast pyrolysis based on the heating rate. Slow pyrolysis occurs when the time taken to heat the biomass in the pyrolysis process is much longer than the pyrolysis reaction time. There are some other alternatives based on the pressure and medium at which pyrolysis process is carried out. Based on the operating characteristics, each pyrolysis process has its own applications and output products. The heating rate at which heat energy is transmitted into the substance determines whether it is fast or slow pyrolysis. Both fast pyrolysis and slow pyrolysis are conducted in an inert atmosphere. Pyrolysis is usually classified into three broad types based on the conditions these processes operate in slow, intermediate or medium, and fast. Product yield and oil characterization from the liquid products of fast pyrolysis and slow pyrolysis of eucalyptus mallee [40] are presented in Table 1.1. Each of these different parameters may lead to variations in the output product. Below, each process is specified in detail.

Table 1.1 Product yields and oil characterization from the liquid products of fast pyrolysis and slow pyrolysis of eucalyptus mallee [40].

	Slow pyrolysis at 420°C	**Slow pyrolysis at 500°C**	**Slow pyrolysis at 600°C**	**Fast pyrolysis at 500°C**
Closure (wt.% on dry feed)	98.9	96.0	95.3	94.4
Heating rate (°C/min)	14.5	13.7	12.6	–
Char yield (wt.% on dry feed)	33.6	31.9	32.0	8.5
Gas yield (wt.% on dry feed)	13.4	11.3	11.4	10.9
Total liquid (wt.% on dry feed)	51.9	52.8	51.9	75.0
(Slow pyrolysis oil from acetone condenser and dry ice)				
Viscosity (cP)	1.4	1.1	1.8	53.8
pH	2.3	2.5	2.4	2.4
Water content (%)	45.9	48.2	43.8	20.8
Char content (%)	0.02	0.03	0.03	0.04
Elemental analysis (%)				
C	28.13	28.00	27.39	43.9
H	9.38	9.38	9.55	7.4
N	0.03	–	–	0.07
O by difference	62.5	62.7	63.1	48.6
HHV (MJ/kg)	6.8	6.4	7.1	14.4
(Slow pyrolysis oil from EP-heavy organic fraction)				
Water content (%)	8.6	11.3	7.8	–
pH	2.8	2.6	2.6	–
Elemental analysis (%)				
C	54.8	53.75	55.30	–
H	7.14	7.41	7.23	–
N	0.10	0.13	0.16	–
O by difference	38.0	38.72	37.32	–
HHV (MJ/kg)	21.3	20.4	21.8	–

HHV, Higher heating value.

1.3.1 Slow pyrolysis

This type of pyrolysis is usually characterized by an extremely slow rate of heating, thereby leading to a low process temperature. The process continues for hours or days due to the high residence time of solids. Slow pyrolysis usually produces more char than liquid product. This technique is preferred in biomasses with high moisture content, as the tolerance toward moisture is very high. It also has the lowest liquid yield among all three types of pyrolysis [41]. Some terms associated with slow pyrolysis are carbonization and torrefaction. To produce coal from biomass, the carbonization process is used, while torrefaction is used to remove highly volatile organic matter from biomass. The resultant torrified biomass is extremely light and insoluble in water and contains an extremely high concentration of energy per unit mass. Torrefaction occurs in a narrow and very low temperature around 200°C–300°C. Meanwhile, carbonization occurs at very high and broad temperature range. Added to this advantage, it has an extended shelf life making it ideal for storage [42–45].

1.3.2 Intermediate pyrolysis

Intermediate pyrolysis takes place in a temperature range of approximately 300°C–500°C. Products of intermediate pyrolysis are characterized by a low tar content and low viscous properties. Although this reaction provides a reduced yield of liquefiable product, the advantage remains that products of intermediate pyrolysis are more versatile in terms of decomposition reactions that may be adopted. As a result, the process may be streamlined as per the needs, with a range of feed sizes being accepted in this type of reaction [35,36,46]. The intermediate pyrolysis leads to very less formation of tar; it is due to controlled chemical reaction instead of heterogeneous chemical reaction. The non-condensable gas residence time is more dependent on reactor type, woody biomass pyrolysis by intermediate pyrolysis generate more liquid fraction around 55% as compared to 75%. However, it is applicable only for woody biomass pyrolysis and it varies with other biogenic feedstocks [47]. Though intermediate pyrolysis process recently overcomes the problem associated with the fast pyrolysis process. The formation of high-level tar and persistent fouling of liquids with higher amounts of char and ash are primary difficulties faced in the quick pyrolysis process, as are biomass conditioning and lower moisture content. Because of the properties of the reaction mixture, which tends to obstruct filters in a short period of time, filters used to limit this normally fail. Slow pyrolysis is also not a viable option because it significantly reduces the organic phase yield and is used to manufacture more solid products like charcoal.

1.3.3 Fast pyrolysis

Fast pyrolysis is the best choice when oil is required to be the desired product. As the name suggests, this reaction adopts a fast decomposition of biomass to produce vapors, which are again rapidly brought to liquid form to reduce the tendency of secondary decomposition. For fast pyrolysis, the vapor residence time is of second or millisecond. Fast pyrolysis primarily yields more bio-oil and gaseous products. This reaction typically makes use of an extremely high heating rate, which necessitates low feed size for improved efficiency. Due to the quick reaction times, operators also have an increased level of supervision over the reaction itself [46,48]. The fast pyrolysis reactors are used recently, which is one type of bubbling fluidized-bed reactor, due to its simple technology with its effective heat transfer because of higher particle density and better temperature control it is well-established. In comparison to a bubbling fluidized-bed pyrolysis reactor, a circulating fluidized-bed pyrolysis reactor is much more complicated. In circulating fluidized-bed reactor, the char residence time is shorter; hence, certain amount of char will form in the pyrolysis oil. To retain this particle, the counter measure will be used to filter. Ablative reactors are a very diverse type of pyrolysis reactor. A large heat transfer surface will be used to achieve the best heat transfer. In the twin screw reactor, the core of the reactor will rotate in same direction, the main advantage of this reactor that it will rotate in same direction due to intertwining flights with heat carrier loop. The heat carrier consists of steel shot or sand [47].

1.4 Mechanism of pyrolysis

Owing to the complex structure of biomass, the pyrolysis process for biomass decomposition is more difficult than other chemical reactions. Various components in biomass will decompose at varying temperature ranges followed by various reaction mechanisms and rates of decomposition. Depolymerization, aromatization, isomerization, dehydration, decarboxylation, and char formation have been identified as parallel and serial reactions in biomass pyrolysis [49]. While heating the biomass, various chemical structures within the biomass are broken, resulting in the formation of volatile compounds and also rearrangement of its matrix structure [50,51].

1.4.1 Mechanism of cellulose pyrolysis

Cellulose is a polymer unit of β-1−4 link D-glucose group. It contains about half of the lignocellulose components. A detailed understanding of decomposition reaction chemistry involving cellulose pyrolysis gives valuable information into the development of effective pyrolysis conversion techniques of waste biomass into biofuels. Cellulose is a key component of biomass's primary cell wall. The cellulose decomposition occurs by the cleavage of glycoside bonds acutely at the temperature of 315°C−400°C [52].

The cellulose depolymerizes into oligosaccharides, which continue to decompose until anhydro-monosaccharides are formed. Levoglucosan developed as an initial product of this breakdown and it will continue to breakdown the isomerization and dehydration reaction to produce anhydro-monosaccharides, for example, dianhydro glucopyranose, anhydro-glucofuranose, and levoglucosenone. These sugars further undergo dehydration reaction to produce furans. Due to decarbonylation and decarboxylation reaction, carbon monoxide and carbon dioxide are formed, while char is produced as a result of polymerization reaction.

The dehydration reaction is the certainly reason for most weight loss in the cellulose structure before 300°C and water molecules escaped from cellulose structure when heated at 200°C [53,54]. Dehydration of intermolecular component results in the development of more covalent bonds that leads to a highest thermal stability and reticulation of polymer. The dehydration of intramolecular structure in cellulose to the production of C = C bonds it will help to form the benzene rings contains char [53].

1.4.2 Mechanism of hemicellulose pyrolysis

In spite of the discrepancy in polysaccharide components of hemicellulose, it thermally breaks down between 200°C and 350°C. The structural similarity between hemicellulose and cellulose is almost similar, but decomposition mechanism of hemicellulose varies as compared to cellulose. During the cellulose pyrolysis, the breakdown of glycosidic link between pyranose rings produces a glucosyl cation, which will help to stabilize the formation of levoglucosan. In other side xylose cation, it was not able to generate any stable anhydride due to the lack of sixth carbon, exchanged oxygen in the fourth position. The xylan decomposition product is acetic, hydroxy acetone, furfural, carbon dioxide, carbon monoxide, and water [55].

Acetic acid is formed when the acetyl group that was bound to the primary xylan chain separates at low temperatures [56]. The pyrolysis behavior of pentoses and hexoses in the hemicellulose structure is different [57]. The hydroxymethyl is the key product in softwood pyrolysis not hardwood hemicellulose mainly comprises of pentoses. Räisänen et al. [58] reported a similar result for the development of hydroxymethyl. According to Patwardhan et al. [59], xylan decomposition follows three major pathways: (1) depolymerization to anhydro sugars and sugars, (2) dehydration to pyran rings and furans, and (3) pyranose and furanose ring breakage to provide lighter oxygenate species. As compared to char yield of cellulose, the char yield of hemicellulose was quite higher. This variation is because the catalytic effect of minerals is more in the hemicellulose as compared to cellulose; this will enhance the formation of char [60,61]. Even after performing the demineralization of hemicellulose, the xylan and glucomannan pyrolysis can provide the yield of char thrice higher as compared to cellulose [50].

1.4.3 Mechanism of lignin pyrolysis

As compared to the cellulose and hemicellulose pyrolysis mechanism, the lignin decomposition mechanism is more complex. The pyrolysis of cellulose reveal that glycoside bond decomposes nearly 350°C and also lignin not showing any sharp and exact decomposition curve and general reaction sequence will be labeled as primary decomposition at 200°C–400°C, and secondary decomposition reaction occurs at a temperature above

400°C. The aromatic methoxy communities are stable while primary pyrolysis stage reaction rate is high at the temperature range of 400°C–450°C. Hence, the aromatic components formed in the primary pyrolysis are primarily 4-subtituted guaiacols and 4-substituted syringols. Mostly, the side chains are unsaturated alkyl communities with a very small quantity of saturated alkyl communities. The important volatile products released from G-lignin at the initial stage include coniferyl aldehyde, coniferyl alcohol, 4-vinyl guaiacol, isoeugenol, vanillin, and acetovanillone [62]. The biomass pyrolysis temperature was increased from 400°C to 450°C, the secondary cracking occurs, and syringols/guaiacols instantly transform into pyrogallols, catechols, and xylenols/*o*-cresols along with phenols. When the temperature raised beyond 700°C, the production of polycyclic aromatic HCs will be raised. Compounds, such as *o*-cresols and phenols, are stable at very high temperature and along with this PAHs also observed.

It is also noted that the reaction related to the methoxyl group will occur at the temperature range of 400°C–450°C and also gasification of catechols occurs at 550°C–600°C [62]. Patwardhan et al. [59,63] reported the production of monomeric compounds with phenol, dimethoxyphenol, 4-vinyl phenol, and 2-methoxy as the major products. The reaction reason for the volatile compound release is primarily due to the instability of propyl chain of bond between monomer units, methoxy constituents of aromatic rings. The releasing of primary volatile leads to the charring process that allows rearranging the char skeleton. In this process, it will release lower molecular noncondensable gases [64]. The pyrolysis oil collected after condensation was rich in oligomeric and dimeric compounds and also observed when temperature increased it will raise the alkylated phenols yield.

1.5 Reactor configurations

1.5.1 Fluidized-bed reactor

The most common type of reactor utilized in case of having an input in solid state is represented in Fig. 1.1. This reactor uses a type of fluidization medium at an extremely high velocity through the bed of the solid-state reactants. This causes the solid to be suspended in the fluid medium and causes it to achieve fluid-like behaviors. Fluidized-bed reactors are favored over other reactor types due to a high and consistent rate of heating, as well as boasting a good command over the reaction and its parameters. As a result, fluidized-bed reactors allow for fast and uninterrupted product production. Another iteration of this technique is the bubbling fluidized-bed reactor. In this method, the fluidizer is usually a gaseous medium at a relatively low velocity. As a result, the bed of biomass is comparatively more stationary than in a standard fluidizing medium. The bed of biomass usually contains certain nonreacting additives, such as sand and reactive catalysts, in a powdered form to promote the reaction [65]. One disadvantage of this method is that the size of input biomass needs to be regulated very vigilantly. Using large biomass fragments may lead to the production

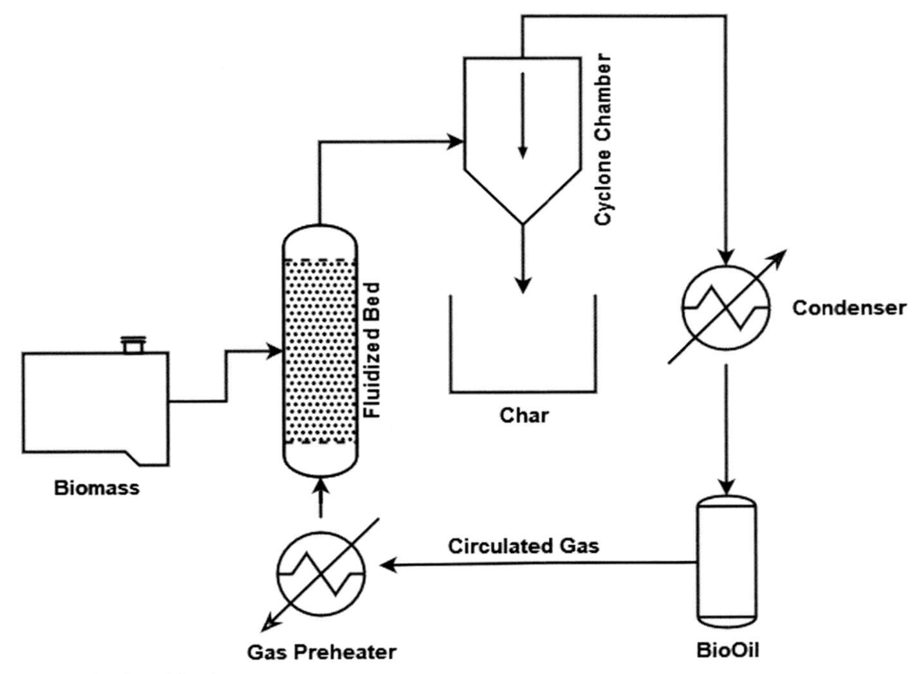

Figure 1.1 Fluidized-bed reactor.

of more char that adheres to reactor walls and is difficult to clean. Similarly, if biomass particle is smaller in size, it may get attached to the fluidizing medium and escape the reactor without undergoing complete pyrolysis reaction.

1.5.2 Circulating fluidized-bed reactor

Circulating fluidized-bed reactors were first introduced by a team of scientists headed by Choi [27], which is shown in Fig. 1.2. They hold the patent for a coiled pyrolysis reactor. The proposed system was different as compared to standard fluidized-bed reactors with the fact that the solid inlet feed is not housed within the reactor, and also on completion of the reaction, resultant char is carried out of the reactor along with the vaporized products. This system also adopts a greater velocity of fluidizing medium, as compared to the standard system [66].

1.5.3 Ablative plate reactor

The characteristic feature of these reactors is that pyrolysis, or decomposition of the inlet feed, is done by bringing the feed in contact with the heated reactor walls as shown in Fig. 1.3. As a result, the reaction speed is directly dependent on the

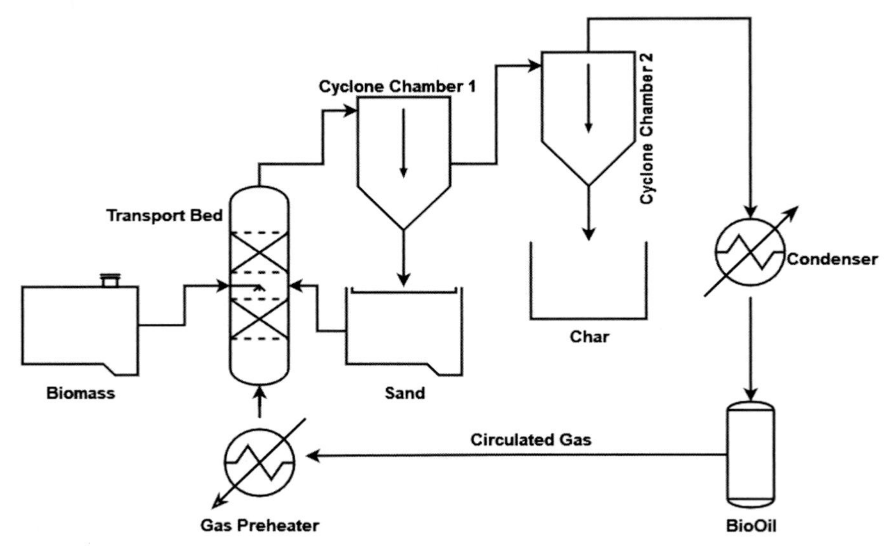

Figure 1.2 Circulating fluidized bed reactor.

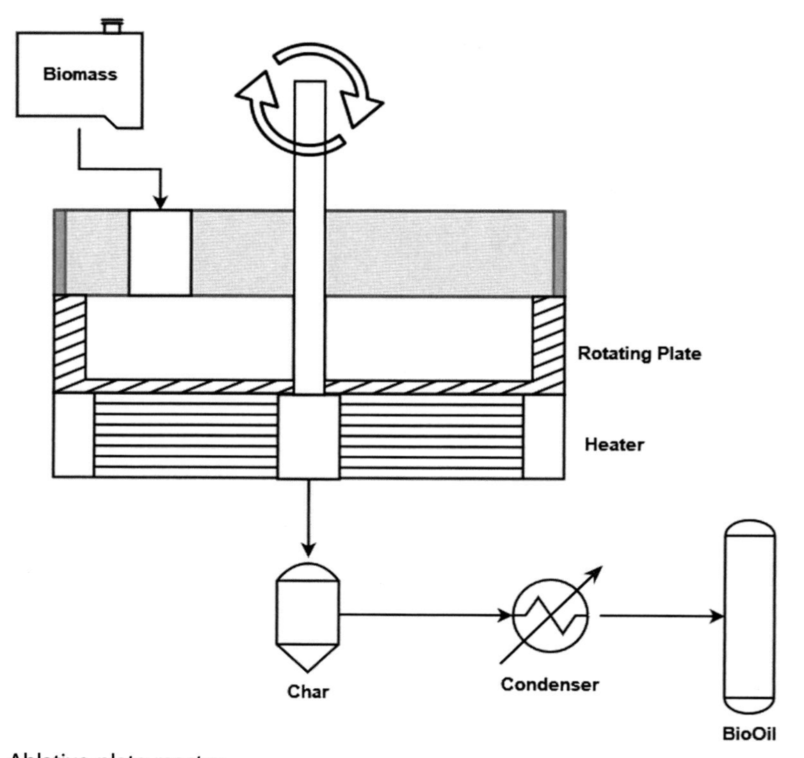

Figure 1.3 Ablative plate reactor.

temperature to which the walls are heated. The reaction rate also depends on ablative pressure, and also on the velocity of biomass relative to the reactor heating surface. The greatest advantage of ablation reaction is that this reaction is more forgiving toward the size of feedstock that is used as input. This is because abrasion causes particle size reduction within the reactor of both the feed and the char [27].

1.5.4 Auger/screw reactor

As the name suggests, the fundamental element of these reactors is a central rotary screw. The rotation of this screw acts as a blender for the inlet biomass and also helps to transport it within the reactor. The rate of reaction is controlled by two major factors—the residence time of feedstock (which is dependent on the speed of rotation of the screw) and the heat energy provided via the reactor walls—and presented in Fig. 1.4. These systems are preferred due to their small sizes, which make them the ideal choice for in situ biomass pyrolysis [67].

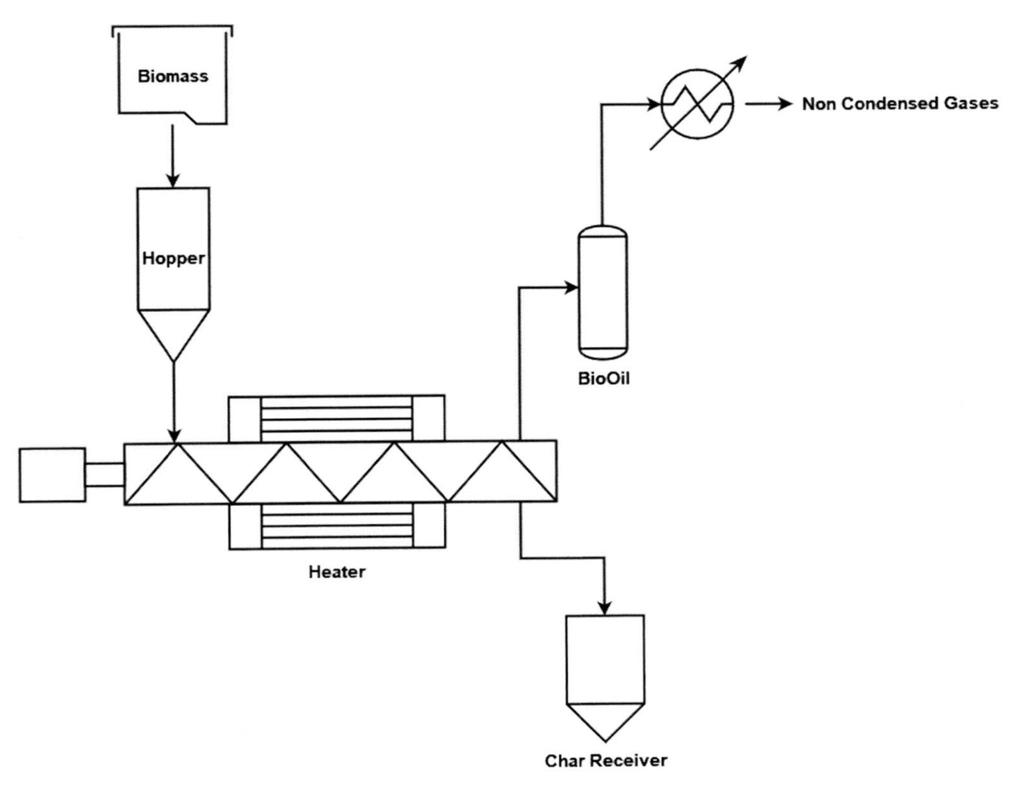

Figure 1.4 Auger/Screw reactor.

1.5.5 Rotating cone reactor

In this system, the two input components, namely biomass and sand, are inserted at the bottom of a spinning conical reactor. The inlet mixture is carried to the top in a helical pattern, due to the centrifugal force caused by rotation. The reason for preferring this system is because it makes use of a gas system to canary the biomass, thereby resulting in a significant cost reduction as compared to the other systems [27]. It was developed in the University of Twente and later brought into commercial usage by BTG-BTL in the Netherlands [18]. Recently, a rotating type reactor having a capacity of up to 50 tons per day was installed in Malaysia and is used to convert bunches of empty fruit from palm trees. Another system in the Netherlands makes use of this technology to pyrolyze residues of wood into bio-oil. Furthermore, both these units recycle the resultant biochar and uncondensed gases by using these as heater fuels for the inlet sand [68].

1.5.6 Cyclone/vortex reactor

This system makes use of a heated and nonreactive gas stream (steam or N_2) to carry the feedstock into the reactor tube and presented in Fig. 1.5. Here, the system makes use of extremely high centrifugal forces to force the reactors to come in contact with the reactor walls. Similar to an ablative plate system, the feedstock is then converted to oil by supplying high amounts of heat to the reactor walls. Any unreacted feedstock particulates are reused by means of a solid recycle system. This system has shown to produce yields of up to 65% and is also proved to be able to handle the requirements of fast pyrolysis reactions [43].

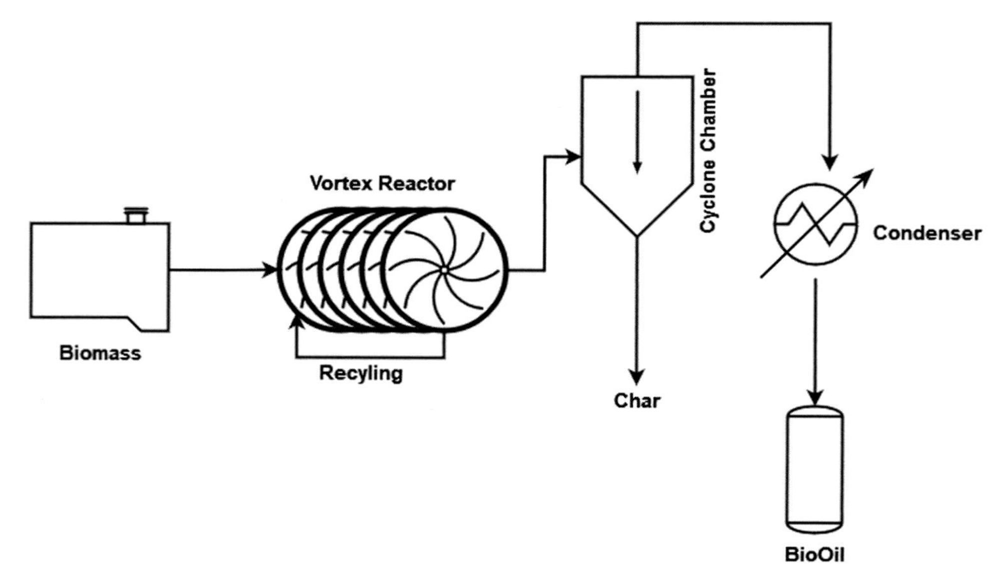

Figure 1.5 Cyclone/vortex reactor.

1.6 Upgradation techniques for pyrolyzed bio-oil

As compared to the baseline diesel fuel, the crude bio-oil obtained after the condensation of volatile matter from pyrolysis is of lower quality. The major parameters that are found to be higher in the obtained bio-oil when compared to diesel are its aqueous/water content, viscosity, acidity, density, oxygen content, and corrosiveness. They also have a different chemical composition than fossil oils, consisting mostly of acids, ketones, aldehydes, esters, phenols, and esters, and are observed to be a convoluted mixture of organic compounds [69]. In addition, the instability caused due to aging and nonmiscibility with HCs is few other properties that prevent us from using the bio-oil directly in a diesel engine and hence requires upgradation or refining. There have been several studies on techniques used in the upgradation of bio-oil with numerous mechanisms. Classification of these methods into three categories, such as physical, chemical, and catalytic, is illustrated in Fig. 1.6.

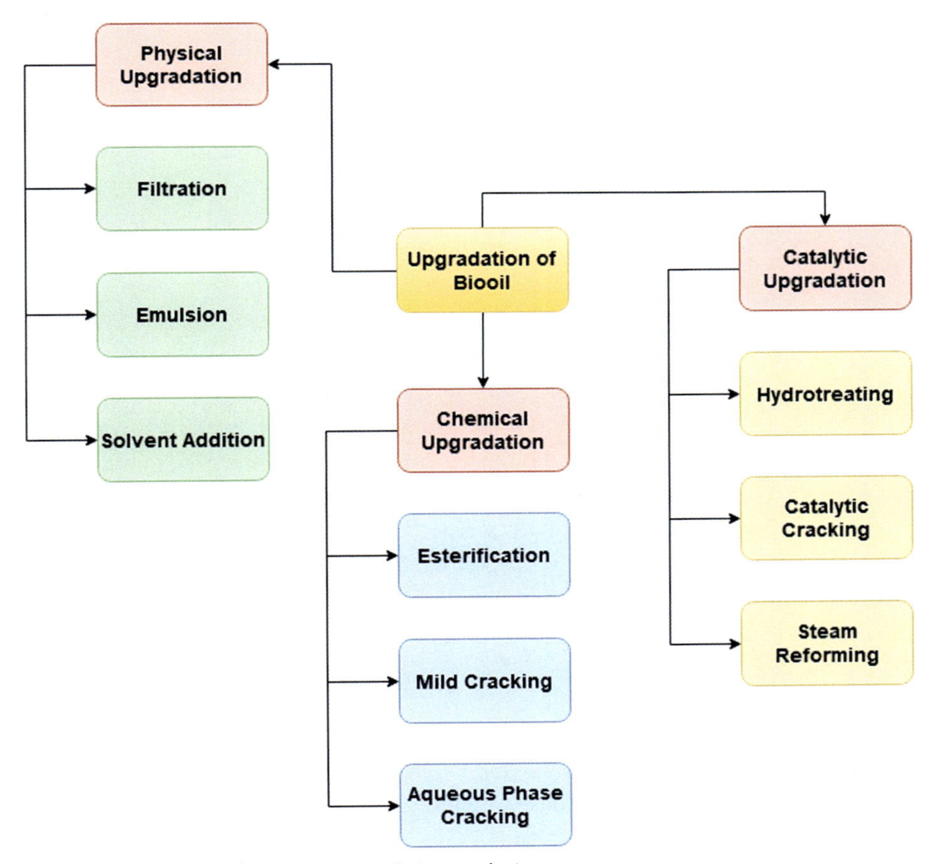

Figure 1.6 Classification of various upgradation techniques.

1.6.1 Physical upgradation of crude bio-oil

Physical upgradation methods are used to overcome bio-oil's incompatibility with conventional fuels due to its higher solids, oxygen, and viscosity material. Three widely known techniques are used to perform the physical upgradation of bio-oil, namely filtration, emulsification, and solvent addition.

1.6.1.1 Hot vapor filtration

This technique is a noncatalytic pyrolysis process and uses filtration materials that can handle the corrosive nature of bio-oil [70]. Since hot vapor filtration (HVF) is effective in reducing alkali (<10 ppm) and ash content (<0.01%), the bio-oil obtained after filtration is of good quality [18]. These levels are significantly lower than the bio-oils that are produced via systems that only used cyclones. It has also been shown to be efficient in lowering the pH of the bio-oil while also producing less biochar. However, this process also minimizes the total yield of the bio-oil (10%–30% of the yield loss in the process) primarily due to the plugging problems that occur in the filter [71,72]. Moreover, it gives biochar that could be catalytically active and leads to a reduction in yield up to 20%. This makes it essential for the self-cleaning of the filter.

Testing of the bio-oil obtained after HVF process in an engine has shown to have a lower ignition delay along with a considerable increase in burning rates than crude bio-oil [73]. Furthermore, it is extremely difficult to filter particles of sizes <5 μm and thus requires very large pressure drops.

1.6.1.2 Emulsification

In general, pyrolysis-derived bio-oil is incompatible with HC-based fuels. However, it has been discovered that these oils can be emulsified with diesel using surfactants, as shown by many research studies [74–77]. For using bio-oil in automobiles, this process offers a simple, short-term solution. Catalytic steam reforming is required first, followed by a water–gas shift reaction in the process. Even though this method requires high energy for production along with reliable and fully developed reactors, they in turn give out hydrogen, which can be used as a clean energy resource. One such method for producing stable emulsions of bio-oil (5%–30%) in diesel fuel has been developed at CANMET [78]. Emulsions containing 5%–95% of bio-oil in diesel have been tested at the University of Florence, Italy [74]. The high cost of surfactant and the huge energy consumption that occurs during emulsification are two major disadvantages of this method.

1.6.1.3 Solvent addition

Two of the main properties that render them not suitable for engine applications are their viscosity and calorific value. As a result, combining the bio-oil with a solvent that has these characteristics is one way to increase heating value while lowering viscosity. Polar solvents, such as ethanol, furfural, and methanol, are the commonly used liquids that shows to be effective in homogenizing the bio-oil, increasing its heating value and reducing the total viscosity [69,79]. The higher heating value (HHV) of the bio-oil increases as these solvents have a higher calorific value than the bio-oil. Several researchers have found that the addition of a solvent immediately after obtaining the bio-oil from pyrolysis exhibits higher stability and homogeneity, in addition to the increased HHV [69]. The addition of a liquid/solvent in the bio-oil has proven in reducing its viscosity via three mechanisms: (1) physically diluting the solvent without modifying the chemical reaction rates; (2) Using molecular dilution or altering the microstructure of bio-oil to minimize the reaction rate; and (3) promoting chemical reactions between the bio-oil and the solvent to stop chain formation. This method is a feasible approach for bio-oil upgradation due to the versatility and low cost of certain solvents, as well as the beneficial effects of the oil properties. Furthermore, several works have been conducted on producing multicomponent mixtures/blends that contain bio-oil [80]. The control over properties of the fuel (flashpoint, viscosity, etc.) is obtained while effectively utilizing the blended biomass-derived liquids.

1.6.2 Chemical upgradation of bio-oil

Three widely known methods are used for the upgradation of bio-oil via chemical methods that include aqueous phase reforming, esterification, and mild cracking.

1.6.2.1 Aqueous phase processing/reforming

This reforming process is among the most promising ways to convert biomass into biofuels, particularly into alkanes or hydrogen. The bio-oil that contains an aqueous phase (mixture of sugars, furfural, anhydro sugar, etc.) is initially made to go under hydrogenation in the presence of RU/C catalyst, and a pressure of 70 bar. As a result, undesirable products, such as methane, diols (C2−C4), sorbitol, and various light gases, are formed. These are then used in the aqueous phase-reforming process, which takes place at pressure 55.1 bar and 265°C in the presence of monometallic catalysts, such as Pt along with Al_2O_3. The end products are H_2 and alkanes [81].

1.6.2.2 Mild Cracking

One of the main disadvantages of using acid-catalyzed cracking (e.g., use of HZSM-5 as a catalyst) or processes, such as hydrotreating (HDT), obtains a large yield of gases, HCs, and coke, which directly plays a role in the performance of the catalyst. Hence, it is necessary to reduce the formation of such products for the efficient functioning of the catalytic material. Fisk et al. [82] have studied the concept of mild cracking reaction and proposed a methodology for this problem. The use of ZnO with calcinated Zn and Mg supported alumina for performing this research at the University of Kentucky. It was found that partial removal of oxygen from the bio-oil was the key to a mild-cracking reaction and would prevent the formation of coke and gases [83].

1.6.2.3 Esterification

The organic acids existing in the crude bio-oil have the tendency to act as catalysts for the polymerization of the sugar and its derivatives (hydroxyl aldehydes or acetones, furfural, etc.) during transportation or storage purposes. Hence, there is a necessity to increase the sustainability of bio-oils by the enhancement of their physiochemical properties. The esterification process is one of the best techniques used in the conversion of the reactive organic acids present in the bio-oil obtained into a stable ester. The reaction occurs in a temperature range between 55°C and 60°C in a time period of 1–4 hour. This heating process involves the use of an acid catalyst (H_2SO_4 solution) that is mixed with a polar solvent.

The elimination of organic acids that is present in the crude bio-oil (formic acid, acetic acid, propionic acids, etc.) can serve a dual purpose: (1) to enhance its stability and (2) to minimize the corrosiveness of the oil. Esterification also has other promising features, such as minimizing the viscosity of the bio-oil and increasing the HHV, volatility, and mixing properties of the bio-oil with diesel fuel. This process can be catalyzed by both acids or bases. However, the usage of base catalysts has the tendency to cause undesirable reactions, such as saponification or neutralization, and thus acid catalysts are preferred [84]. Even though the bio-oil contains acids inherently, the weak nature of these acids makes it unsuitable to be used as intrinsic catalysts for esterification [85]. Hence, homogenous acid catalysts, such as hydrochloric acid or sulfuric acid, are often used for esterification. These catalysts offer high product yields, but at the same time it is difficult to be separated during catalyst recovery. Hence, the esterification process is also carried out with heterogeneous catalysts, such as sulfated metal oxides [86], acid-functionalized porous silica [87,88], and ion-exchange resin [87,89]. These heterogeneous catalysts overcome the drawbacks that are offered by using homogeneous catalysts. An acid catalyst system that contains zeolites (HZSM-5, HY, Hbeta, etc.) is among the most used catalysts for bio-oil upgradation for three reasons: (1) strong acidic sites, (2) large surface area, and (3) good thermal stability.

Using Amberlite IR-120 and Amberlyst 15 ion-exchange membranes, Wang et al. studied the catalytic esterification of bio-oil upgradation. The result shows that bio-oil acid number decreased by 85.95% and 88.5%. As a result, a significant number of acids would be converted to neutral esters. Also, calorific value of pyrolysis oil was improved by 31.64% to 32.26%, respectively, and the water content was reduced by 27.74% from 30.87% [90]. The component of esterified pyrolysis oil was examined using gas chromatography and Fourier transform infrared spectroscopy in a previous study, and the results show that phenolic content in residue is the highest, followed by HCs and ketones. Esterification of bio-oil also increases the organic fraction and fully removes the formation of char [91]. The one-step esterification hydrogenation method of bio-oil was investigated by Xu et al. Aldehyde and ketone contents, as well as organic acid concentration, were found to have decreased from 28% to 0% and 49% to 8%, respectively. They found that one-step esterification-hydrogenation could be an effective way to increase the light fraction of bio-oil because the total esters and alcohol content were 87.24%. The esterification and ozone oxidation process decreased the water content from 44.75% to 2.38% and increased the heating value from 9.5 to 25 MJ/kg [92].

Peng et al. compared catalytic upgrading processes in subcritical and supercritical ethanol, observing that the supercritical upgradation process was more successful than the subcritical upgradation process. Water is the primary by-product of this reaction; the esterification process is reversible and it will be carried out using a reactive distillation column to remove and separate water during the reaction to increase the reaction yield [91,93]. For the esterification process, homogeneous and heterogeneous catalysts will be used. Moreover, for bio-oil upgradation, heterogeneous catalyst will be preferred because the separation of catalyst in the product is very easy. Aluminum silicate and HZSM-5 are the best catalysts for bio-oil upgradation [93,94]. The esterification process was reported to enhance the bio-oil quality that includes a decrease in water content, oxygen content, aging rate, acidity, density, and viscosity and an increase in calorific value. However, due to simplicity, the low pressure, and temperature, low-cost alcohols, such as esterification by methanol, seem to be a promising technique to enhance the quality of bio-oil. In addition to an appropriate catalyst, the choice of alcohol plays an important role during this process. Methanol is the most commonly used alcohol for the esterification process due to its high reactivity with the organic acids that are found in the bio-oil. However, the process must take place at a temperature lesser than 100°C to prevent coke formation. Furthermore, since water makes up a large portion of the crude bio-oil, it comes out as a product during the esterification process, and hence water separation methodologies are vital for this process.

1.6.3 Catalytical upgradation of bio-oil

The use of catalysts in the upgradation of bio-oil is done via three different methods, namely HDT, steam reforming, and catalytic cracking.

1.6.3.1 Hydrotreating

Generally, petroleum products (fuels in particular) that contain a large quantity of hydrogen are recognized as fuels of high quality. On this basis, one such process named hydrogenation has been used in refineries for the upgradation of fuels. This process, when used on petroleum fuels or bio-oils, has been referred to as HDT. The HDT process is a nondestructive hydrogenation method that is used to improve fuel quality with minimal changes to the boiling range of the fuel. The reaction usually takes place under moderate conditions (~500°C and 20 MPa pressure) while having a constant supply of hydrogen. In particular, HDT is used to displace the oxygen content in the crude bio-oil as water in reacting with H_2. CoMo and NiMo were the two most widely used catalysts and were supported on alumina for this process. It was found that NiMo catalyst gives better results (more catalytic activity) when compared to the other [95]. However, various problem also occured while using these catalysts, such as: (1) instability of alumina or aluminosilicate supports used by catalysts in high H_2O level environments of the oil, (2) necessity for constant resulfurization, and (3) rapid formation of coke which hindered catalytic activity. Hence, there has been a shift toward using precious metal catalysts for the same that are supported on mildly susceptible catalysts in recent years. The major advantage of this process is its cost-effectiveness and is also the most common process that is used in recent/modern refineries due to its commercialization worldwide. However, there is high coking (8%–25%) present post the process and the fuel obtained loses its quality due to it. This also promotes catalyst deactivation due to the poor C/H ratio.

1.6.3.2 Catalytic cracking

This technique can be regarded as one of the most promising methods for the upgradation of bio-oil into liquid fuel. The main advantage of this technique over HDT is that it can remove oxygen from the bio-oil in the form of CO_2, CO, and H_2O and does not require H_2 to do so [96]. This method can be broadly classified as OFF-line cracking and ON-line cracking. In the OFF-line cracking, the bio-oil itself is used as the raw material, whereas in the ON-line cracking, the vapors obtained from the pyrolysis process are used as raw materials [20]. There is a wide range of light products that are obtained through this method due to the combination of catalytic reactions along with hydrogenation and the multiplicity of reactions that occur in the reactor. This process is mostly carried out in the presence of a dual-function catalyst. Zeolites generally offer the cracking function, tungsten oxides or platinum are used to catalyze reactions, and nickel is commonly used to provide hydrogenation function. Alumina is the support that is highly used for this process.

The catalytic cracking done for bio-oil upgradation involves different reactions in the reactor, such as cracking, hydro cracking, hydrogenation, decarbonylation, and decarboxylation, which are previously reported for zeolite cracking and hydrodeoxygenation due to more compound diversity in the bio-oil [97]. The crude pyrolysis oil was enhanced with the catalyst, namely HZSM-5 in super- and subcritical conditions, and it is found that the amount of esters was increased while the quantity of acids was decreased [91]. Hydrodeoxygenation is a catalytic cracking process that uses high pressure hydrogen and a catalyst to remove oxygen. Various forms of oxygenated groups, such as esters, ketones, phenols, aldehydes, and acids, will be converted into carbon dioxide and water in this process, which will decrease the oxygen percentage [98,99]. Hong et al. [100] suggested Pt/HY, which is a bifunctional catalyst as a good choice for the hydrodeoxygenation of phenolic compounds. For guaiacol hydro-conversion, Nimmanwudipong et al. used Pt/Al_2O_3 as a hydrodeoxygenation catalyst and Zhu et al. used Pt/HBeta of anisole as a hydrodeoxygenation catalyst and this catalyst improved the hydrodeoxygenation [101]. Cheng et al. [102] performed catalytic cracking using bimetallic catalyst of $Fe\text{-}Co/SiO_2$ for a pyrolysis oil and found that bio-oil HC concentration was increased. White et al. tested $NiMo/Al_2O_3$ for a bio-oil derived from coal at a pressure of 20.7 MPa of hydrogen and a temperature range of 400–500°C and found that in the bio-oil oxygen concentration varies linearly with the conversion rate. At 360°C and 2 MPa, the upgradation behavior of the bio-oil was investigated by Zhang et al., in the presence of catalyst Co-Mo-P/Al_2O_3 and hydrogen solvent as a tetralin solvent. The oxygen content of bio-oil was found to have reduced from 41.8% to 3% [103]. According to Fisk et al. [104], the oxygen content of bio-oil was reduced from 41.4% to 2.8% using Pt/Al_2O_3 as a result of this reaction; the aromatic compounds in the bio-oil were increased.

Romero et al. studied the effect of MoS_2 catalyst and recommended a reaction mechanism. The bio-oil oxygen molecules adsorbed on the active site of MoS_2 resulting in compound activation will also present along with catalyst edge as these produced from hydrogen exist in the feed. This allows donation of proton to a bonded molecule from sulfur, resulting in the carbocation reaction. It will undergo straight breaking of C–O bonds. This ultimately produces oxygen and deoxygenated compounds; henceforth, it removes water formation [105]. Corma et al. studied the mechanism of zeolite cracking and found that cracking mechanisms are primarily dependent on series of reactions. The HC molecules get converted into short fragments through general cracking reaction. Oxygen elimination is associated with decarbonylation, decarboxylation, and dehydration [106]. Twaiq et al. [107] studied that the conversion of palm oil into gasoline by using catalyst, such as composite HZSM-5-coated mesoporous crystalline material, was found to produce gasoline yield around 47%. Adjaye and Bakhshi [108] found that aromatic HC will be enhanced by using zeolite catalyst, such as H-mordenite catalyst and HZSM-5. Judit et al. studied the biomass derivatives from fast pyrolysis via in situ catalytic process and found that the yield of HC increases in organic phase while the acid content was decreased [109]. Zhang et al. [15] demonstrated that zeolite cracking is not proper route for bio-oil upgradation as compared to hydrodeoxygenation. Shakya et al. [110] investigated the bio-oil catalytic upgrading using five various types of catalyst and found that around 35%–40% of produced bio-oil possess similar property like diesel fuel. Studies conducted by Koike et al. [111] shows that catalytic cracking in pyrolysis process reduces the bio-oil oxygen content by about 0.5% of the initial value and upgraded pyrolysis oil were primarily consist of phenolic compounds.

1.6.3.3 Hydrodeoxygenation

The hydrodeoxygenation process is conducted in a hydrogen atmosphere in the presence of catalysts. This environment removes oxygen present in the pyrolysis oil in the water form. Hydrodeoxygenation process widely studied in the past few years [112,113]. Hydrodeoxygenation of pyrolysis oil will be carried out in a wide range of temperature and pressure that varies typically in the range of 200°C–350°C and 50–150 bar, respectively Wang et al. [114] recently reviewed the process parameter, reaction mechanism, and limitations of hydroprocessing process. The hydrodeoxygenized pyrolysis shows the oxygen content of less than 20 wt.%; also, it depends on the method of deoxygenation treatment. Normally for higher degree of hydrodeoxygenized pyrolysis, oil will be formed primarily by more HCs; hence, it directly blends pyrolysis oil with fossil fuel. The main factor determining the economics of hydrodeoxygenation of pyrolysis oil is the operating condition and hydrogen consumption rate and pressure of the particular reaction. Pyrolysis oil contains higher amount of aromatic compounds particularly phenolics, a higher feature is the development of catalyst for hydrodeoxygenation that promotes various oxygen-contained functionalities without aromatic ring hydrogenation it implies more consumption of hydrogen [114,115]. Hydrodeoxygenation states have been assessed depending on phases, physical state, and origin of pyrolysis oil. The hydrodeoxygenation can apply both aqueous and organic phases present in the pyrolysis oil.

The oxygen molecules present in the bio-oil can release in the form of water during the hydrodeoxygenation process through a series of C—C, C—O—C, and C—OH bond cleavage and deoxygenation reactions [116]. Once the hydrodeoxygenation process was over, the bio-oil will be upgraded into energy-rich and noncorrosive liquid fuel, for example, naphtha, with less oxygen content 2 wt.% [117]. However, cost associated with the hydrogen consumption is one of the main obstacles in using hydrodeoxygenation process because it requires very high partial pressure supply of hydrogen. To address these issues, mild hydrodeoxygenation would effectively minimize the hydrogen consumption and catalyst costs, making pyrolysis-based bio-oil a better option than the method used in petroleum refineries. The key technique in this is to split the oxygen removal process into two stages: first, the raw bio-oil needs to stabilize at low temperature around 100°C—280°C and then the product is put into the secondary reactor and is maintained at high temperature around 350°C—400°C for the additional removal of oxygen [118—120]. French et al. investigated the behavior of catalysts, such as $NiMo/Al_2O_3$, in the HDT of bio-oil for upgradation. In the first step, combining catalyst and bio-oil in the reactor chamber reactant allowed for 1-hour stabilization at 1000 psig and 280°C. Furthermore, the reaction chamber was heated to a temperature of 400°C and was maintained this temperature for 1 hour. The results at this condition show that the product oxygen content was decreased to 6.7 wt.% and hydrodeoxygenation in this condition was an economical and feasible technique for eliminating the water-soluble and water compounds, such as carboxylic acids. This result recommended that hydrodeoxygenation gives an alternative route to stabilize products that will accommodate in already existing refinery system for processing [121]. Recently hybrid one-step hydrodeoxygenation was introduced to stabilize carboxylic acids and aldehydes, which are the two primary functional groups in bio-oil [122]. According to Tang et al. [123,124], arenesulfonic acids have higher catalytic activity than propylsulfonic acids because the acid strength is higher.

1.6.3.4 Steam reforming

Initially, the word reforming was coined to illustrate the conversion of the petroleum fractions with thermal energy to produce higher volatile products having greater octane numbers. It also represented the net effect due to simultaneous reactions that include dehydrogenation, isomerization, and cracking [125]. However, it is now used to refer to the transformation of gaseous HCs and volatile organic matter into H_2 containing gases, such as syngas (mix of CO and H_2). Steam reforming of the bio-oil is a widely researched process as it produces an upgraded bio-oil along with a clean and clear H_2. This H_2 can be used both as a raw material for industrial application and also in fuel cells for power generation. The process is carried out at high temperatures (800°C–900°C) [126] in the presence of a catalyst. Ni catalyst is shown to demonstrate excellent activity for processing bio-oil [127]. The coke formation that begins at a temperature range of 575°C–900°C can sometimes deactivate the catalyst [128]. The process begins with the fast pyrolysis of biomass and converts the oil into hydrogen. This step is also sometimes accompanied by a shift conversion [129]. Sequential condensation or addition of water can be done to separate the bio-oil into two phases—a hydrophobic phase that mainly constitutes lignin-derived oligomers, and a water-rich phase that is predominantly composed of carbohydrate-derived compounds. The major advantage of this process is that the production of such coproducts improves the overall economy of the process. The water-rich phase can undergo steam reforming to produce hydrogen, while the rich lignin-oligomer found in the hydrophobic phase can be used as an alternative for phenol in phenol-formaldehyde resins.

The bigger reactor unit till now used in China is a fixed bed two-stage reactor system with the inner diameter of 20 mm and the height of 800 mm [130]. To achieve better steam-reforming effect, it is necessary to maintain the system at very high temperature around 800°C–900°C [131]. Chornet et al. is the first researcher who studied the steam-reforming process of bio-oil aqueous portion and proposed a combined hydrogen production concept through aqueous fraction of bio-oil. They have primarily investigated the effect of Ni-based catalysts [132]. Basagiannis and Verykios studied the effect of Ni and noble metal-supported types, such as Al_2O_3, MgO/Al_2O_3, and La_2O_3, for the purpose of bio-oil reforming [133]. Hu and Lu reveal that Ni–Co-supported catalyst is considered as the best catalyst pyrolysis oil reforming to produce hydrogen at the temperature of 673 K with an S/C ratio of 7.5:1. This catalyst demonstrates higher activity and very good stability to produce CO-free hydrogen [134]. Hydrogen generation by steam-reforming process of bio-oil combined with in situ CO_2 capture was studied. Maximum yield of hydrogen was found 75% in the existence of calcined dolomite [135]. Takanabe et al. investigated the steam-reforming process of acetic acid over the surface of Pt/ZrO_2 catalyst. Taknabe et al. [135] used dual functional mechanism for steam reforming in acetic acid for, both ZrO_2 and Pt where participated in the reaction. The existence of platinum is essential for the production of hydrogen and then water is activated on zirconium dioxide to form an extra surface hydroxyl groups that can participate in water–gas shift reaction and steam reforming [136]. Similarly, Yamazaki and Matsuki [137] investigated the nanoporous catalyst Ru/ZrO_2 by sol–gel technique and template technique for steam reforming of pyrolysis oil. Cyrille et al. investigated the steam-reforming model compounds of pyrolysis oil, such as phenols, acetic acid, ethanol, and acetone, over Pt, Rh, and Pd impregnated on alumina. They found that the material of ceria–zirconia showed more activity than alumina sample [138]. Chen et al. studied the carbon deposition characteristics in the reforming process of pyrolysis oil for the production of hydrogen and found that the carbon elimination and deposition and the highest value of coke formation were achieved in a wide range of temperature 575°C–900°C; at the same time higher percentage of steam to carbon results in the elimination of carbon. Also, catalyst regeneration shows minor drop-in activity because of iron contamination and nickel redispersion [112].

1.6.3.5 Supercritical fluids

Fluids that present at a pressure and temperature greater than its critical point but lower than a pressure that could condense it back to a liquid are called supercritical fluids (SCFs). Being in a supercritical state makes these fluids exhibit properties that digress from the normal. It can also be described as a homogeneous fluid phase that exists between that of a liquid and a gas. Particularly, they possess the ability to dissolve materials like liquids and effuse like gas through solid materials. SCF's specific transport properties facilitate liquefaction and gasification reactions by allowing compounds that are normally insoluble in the solvent's gas or liquid phases to be dissolved [139]. The diffusivity and viscosity of an SCF resemble that of a gas, while its density, on the other hand, is similar to that of a liquid. Usage of SCF for upgradation has shown to be effective in increasing the HHV and reducing the viscosity of crude bio-oils. Although water is most commonly used as the SCF for HDT processes, using it in the form of a solvent during the liquefaction process of biomass has two major disadvantages: (1) oil-product yield (insoluble in water) is reduced and (2) the obtained bio-oil has a high oxygen content and viscosity. Several studies have reported an increased oil yield by using organic solvents, such as butanol, methanol [140,141], ethanol [140,142], and acetones [143,144]. The use of SCF is preferred when compared to the catalytic upgradation process as there are no requirements for a constant external supply of material like H_2. Moreover, this process does not produce coke, which has a tendency for catalytic deactivation. Thus SCFs could be used in the upgradation of bio-oil without the presence of catalysts. SCFs are also shown to demonstrate catalytic like activities during the upgradation process, which is larger than the observed activity in general liquid–solvent conditions. Though batch reactors are used in several studies for upgrading bio-oil using SCF, continuous flow reactors are a better alternative that could offer an increased production efficiency and is also more suitable to be employed commercially [145].

Supercritical fluids are not only utilized for reaction condition to generate bio-oil but also used as very good medium to improve the quality of bio-oils, and they have greater potential to produce bio-oil with lesser viscosity and more heating value [145]. Usually, supercritical fluid treatment for improving bio-oil quality will use some external catalysts, such as HZSM-5, aluminum silicate, and bifunctional catalysts [112]. The bio-oil upgradation process was mainly conducted in autoclave reactor volume of 150 mL. Once upgraded bio-oil components were optimized and the bio-oil properties were enhanced greatly. In supercritical media, the catalyst can promote the conversion of majority of acids into different sorts of esters in the bio-oil upgradation process. SCF treatment results in the reduction of kinematic viscosity and density of bio-oil, at the same time pH and heating value will be increased to a certain value [146]. Dang et al. found that high pressure of hydrogen around 2.0 MPa will effectively eliminate the coke formation. Though temperature increment was helpful to promote the calorific value of upgraded oil and the quantity of useful product decreased and coke formation will be the more serious problem. The process of supercritical fluid upgradation of bio-oil is environmentally sound and can be conducted at comparatively low temperature. The supercritical fluid treatment is not cost-effective because of organic solvents used here are high cost [69,147]. The oil extraction pressure and temperature play an important role in the solvability of element in solvent, which primarily based on the chemical properties of target compounds, such as lutein and astaxanthin, shows thermal degradation at very high temperature; hence, their range needs to be optimized for production [148,149]. In supercritical fluids, the CO_2 production of bioactive compounds and the efficiency of the extraction increase with carbon dioxide temperature and pressure up to optimum level. However, high temperature causes the compound thermal degradation, while more pressure will block the diffusivity of fluid matrix, which results in extraction yield reduction [150]. Also, higher pressure and temperature imply the formation of extra waxes. Temperature and pressure impact the solvation power of supercritical fluid because of their response to the density of solvent and it can be modified by operating temperature and pressure. However, increasing pressure and constant temperature cause an increase in the solvation power of oil and carbon monoxide density; also, it increases the solubility of extraction yield and bioactive compounds. Hence, solubility is more related to the supercritical fluid to carbon dioxide density and solute properties, such as polarity, molecular mass, and vapor pressure [150–152].

1.7 Energy recovery for heating or process applications

Pyrolysis oil will be substituted for the petroleum or diesel fuel process for many stationary applications that include furnace, boiler for heat generation and gas turbine, and diesel engine for electricity generation. Many investigations reported that pyrolysis oil can be blend with diesel fuel in the range of 5%–30% and is no big problem while running engine with B20 (80% diesel with 20% pyrolysis oil blend by volume) [153]. The first research work on utilizing biomass pyrolysis liquid fuel in diesel engine was conducted by VTT in Finland. Performance and emission characteristics of engine were conducted in a 4.8-kW test engine [154]. Power generated by biomass pyrolysis has been evaluated as a leading technology for minimizing carbon dioxide emission in power production [155]. The very important industrial-based alternative nowadays is Rankine cycle by using waste biomass. The power producing plants less than 1 MWe and up to 240 MWe is currently in operation by utilizing solid biomass as a fuel. Some other power plant fueled with biologically derived fuel, such as anaerobic gases, landfill gases, and gasification-produced fuel gases. Pyrolysis oil can be directly substituted for combustion purpose in furnace or boiler and it can be utilized to produce heat. For pyrolysis oil, this is a simple way to use it directly for application. In this process, natural gas, heavy fuel oil was replaced with the pyrolysis oil; hence, this will reduce the carbon dioxide emission by nearly 90%. [156]. Different manufacturers nowadays assured the performance when utilizing the pyrolysis oil in their burner, for example, Dreizler from Germany and Stork Therme from Netherland. Hence, the pyrolysis oil will be mixed with heating oil/natural gas and storage tank, pipes need to be equipped from stainless steel. By substituting natural gas with pyrolysis oil from Empyro plant for milk powder production company, the carbon emission has reduced by 15%. Every year the natural gas of 10 million m^3 was saved, which is almost equaled 8,000 Dutch household consumption [157].

1.8 Conclusion

Bioenergy plays a crucial role in reducing greenhouse gas emission. Various technologies are proposed to convert waste biomass into useful form of product, such as biofuels for energy applications. Biomass pyrolysis is a promising technique for converting biomass into liquid fuel and value-added chemicals that can be used as a replacement for vehicle fuel and industrial chemicals. To use the method on a large scale, it is necessary to research the composition of biomass as well as the decomposition mechanism, decomposition behavior, and product chemistry. Various types of pyrolysis reactors have been explored to take advantage of the biomass pyrolysis process on a larger scale. Due to higher oxygen content and poor stability, the produced pyrolysis oil is unsuitable for the use of transportation fuel. Pyrolysis oil obtained from different feedstock can be upgraded by physical and chemical routes for the substitution of petroleum fuel. Pyrolysis oil produced through catalytic cracking, hydrodeoxygenation, esterification, steam reforming, and supercritical fluids will produce clean, renewable hydrogen with better calorific value, which makes notable impact on power generation and transportation fuels.

References

[1] Liu C, Wang H, Karim AM, Sun J, Wang Y. Catalytic fast pyrolysis of lignocellulosic biomass. Chem Soc Rev 2014;43:7594–623. Available from: https://doi.org/10.1039/c3cs60414d.

[2] Kalogirou SA. Solar thermal collectors and applications. Prog Energy Combust Sci 30; 2004. < https://doi.org/10.1016/j.pecs.2004.02.001 >.

[3] Serrano-Ruiz JC, Dumesic JA. Catalytic routes for the conversion of biomass into liquid hydrocarbon transportation fuels. Energy Environ Sci 2011;4:83–99. Available from: https://doi.org/10.1039/c0ee00436g.

[4] Abnisa F, Wan Daud WMA. A review on co-pyrolysis of biomass: An optional technique to obtain a high-grade pyrolysis oil. Energy Convers Manag 2014. Available from: https://doi.org/10.1016/j.enconman.2014.07.007.

[5] Manzano-Agugliaro F, Alcayde A, Montoya FG, Zapata-Sierra A, Gil C. Scientific production of renewable energies worldwide: an overview. Renew Sustain Energy Rev 2013;18:134–43. Available from: https://doi.org/10.1016/j.rser.2012.10.020.

[6] Alonso DM, Wettstein SG, Dumesic JA. Bimetallic catalysts for upgrading of biomass to fuels and chemicals. Chem Soc Rev 2012;41:8075–98. Available from: https://doi.org/10.1039/c2cs35188a.

[7] Regalbuto JR. Engineering: cellulosic biofuels—got gasoline? Science 2009;325:822–4. Available from: https://doi.org/10.1126/science.1174581.

[8] Zhou CH, Xia X, Lin CX, Tong DS, Beltramini J. Catalytic conversion of lignocellulosic biomass to fine chemicals and fuels. Chem Soc Rev 2011;40:5588–617. Available from: https://doi.org/10.1039/c1cs15124j.

[9] Isikgor FH, Becer CR. Lignocellulosic biomass: a sustainable platform for the production of bio-based chemicals and polymers. Polym Chem 2015;6:4497–559. Available from: https://doi.org/10.1039/c5py00263j.

[10] Leibbrandt NH, Knoetze JH, Görgens JF. Comparing biological and thermochemical processing of sugarcane bagasse: an energy balance perspective. Biomass Bioenergy 2011;35:2117–26. Available from: https://doi.org/10.1016/j.biombioe.2011.02.017.

[11] Brown TR, Thilakaratne R, Brown RC, Hu G. Techno-economic analysis of biomass to transportation fuels and electricity via fast pyrolysis and hydroprocessing. Fuel 2013;106:463–9. Available from: https://doi.org/10.1016/j.fuel.2012.11.029.

[12] Huber GW, Iborra S, Corma A. Synthesis of transportation fuels from biomass: Chemistry, catalysts, and engineering. Chem Rev 2006;106:4044–98. Available from: https://doi.org/10.1021/cr068360d.

[13] Mortensen PM, Grunwaldt JD, Jensen PA, Knudsen KG, Jensen AD. A review of catalytic upgrading of bio-oil to engine fuels. Appl Catal A Gen 2011;407:1–19. Available from: https://doi.org/10.1016/j.apcata.2011.08.046.

[14] Czernik S, Bridgwater AV. Overview of applications of biomass fast pyrolysis oil. Energy Fuels 2004;18:590–8. Available from: https://doi.org/10.1021/ef034067u.

[15] Zhang Q, Chang J, Wang T, Xu Y. Review of biomass pyrolysis oil properties and upgrading research. Energy Convers Manag 2007;48:87–92. Available from: https://doi.org/10.1016/j.enconman.2006.05.010.

[16] Taarning E, Osmundsen CM, Yang X, Voss B, Andersen SI, Christensen CH. Zeolite-catalyzed biomass conversion to fuels and chemicals. Energy Environ Sci 2011;4:793–804. Available from: https://doi.org/10.1039/c004518g.

[17] Elliott DC, Hart TR, Neuenschwander GG, Rotness LJ, Olarte MV, Zacher AH, et al. Catalytic hydroprocessing of fast pyrolysis bio-oil from pine sawdust. Energy Fuels 2012;26:3891–6. Available from: https://doi.org/10.1021/ef3004587.

[18] Bridgwater AV. Review of fast pyrolysis of biomass and product upgrading. Biomass Bioenergy 2012;38:68–94. Available from: https://doi.org/10.1016/j.biombioe.2011.01.048.

[19] Vispute TP, Zhang H, Sanna A, Xiao R, Huber GW. Renewable chemical commodity feedstocks from integrated catalytic processing of pyrolysis oils. Science 2010;330:1222–7. Available from: https://doi.org/10.1126/science.1194218.

[20] Hew KL, Tamidi AM, Yusup S, Lee KT, Ahmad MM. Catalytic cracking of bio-oil to organic liquid product (OLP). Bioresour Technol 2010;101:8855–8. Available from: https://doi.org/10.1016/j.biortech.2010.05.036.

[21] Elliott DC. Historical developments in hydroprocessing bio-oils. Energy Fuels 2007;21:1792–815. Available from: https://doi.org/10.1021/ef070044u.
[22] Isahak WNRW, Hisham MWM, Yarmo MA, Yun Hin TY. A review on bio-oil production from biomass by using pyrolysis method. Renew Sustain Energy Rev 2012. Available from: https://doi.org/10.1016/j.rser.2012.05.039.
[23] Zheng Y, Zhao J, Xu F, Li Y. Pretreatment of lignocellulosic biomass for enhanced biogas production. Prog Energy Combust Sci 2014. Available from: https://doi.org/10.1016/j.pecs.2014.01.001.
[24] Klemm D, Heublein B, Fink HP, Bohn A. Cellulose: fascinating biopolymer and sustainable raw material. Angew Chem Int Ed 2005. Available from: https://doi.org/10.1002/anie.200460587.
[25] Zakzeski J, Bruijnincx PCA, Jongerius AL, Weckhuysen BM. The catalytic valorization of lignin for the production of renewable chemicals. Chem Rev 2010. Available from: https://doi.org/10.1021/cr900354u.
[26] Calvo-Flores FG, Dobado JA. Lignin as renewable raw material. ChemSusChem 2010. Available from: https://doi.org/10.1002/cssc.201000157.
[27] Dhyani V, Bhaskar T. A comprehensive review on the pyrolysis of lignocellulosic biomass. Renew Energy 2018. Available from: https://doi.org/10.1016/j.renene.2017.04.035.
[28] Harmsen P, Bermudez L, Bakker R. Literature review of physical and chemical pretreatment processes for lignocellulosic biomass. Biomass; 2010.
[29] Laureano-Perez L, Teymouri F, Alizadeh H, Dale BE. Understanding factors that limit enzymatic hydrolysis of biomass: characterization of pretreated corn stover. Appl Biochem Biotechnol Pt A Enzyme Eng Biotechnol 2005. Available from: https://doi.org/10.1385/ABAB:124:1-3:1081.
[30] Collard FX, Blin J. A review on pyrolysis of biomass constituents: mechanisms and composition of the products obtained from the conversion of cellulose, hemicelluloses and lignin. Renew Sustain Energy Rev 2014. Available from: https://doi.org/10.1016/j.rser.2014.06.013.
[31] Lee H v, Hamid SBA, Zain SK. Conversion of lignocellulosic biomass to nanocellulose: structure and chemical process. Sci World J 2014. Available from: https://doi.org/10.1155/2014/631013.
[32] Wang S, Dai G, Yang H, Luo Z. Lignocellulosic biomass pyrolysis mechanism: a state-of-the-art review. Prog Energy Combust Sci 2017. Available from: https://doi.org/10.1016/j.pecs.2017.05.004.
[33] Wheeler E.A. Methods in Lignin Chemistry. S.Y. Lin, C.W. Dence (eds.), 578 pp., illus., 1992. Springer Series in Wood Science. Springer Verlag, Berlin, Heidelberg, etc. ISBN 3-540-50295-5. Price DM 480.00 (hardcover). IAWA J 2014. <https://doi.org/10.1163/22941932-90001308>.
[34] Awasthi A, Dhyani V, Biswas B, Kumar J, Bhaskar T. Production of phenolic compounds using waste coir pith: estimation of kinetic and thermodynamic parameters. Bioresour Technol 2019. Available from: https://doi.org/10.1016/j.biortech.2018.11.073.
[35] Dhyani V, Awasthi A, Kumar J, Bhaskar T. Pyrolysis of Sorghum straw: effect of temperature and reaction environment on the product behavior. J Energy Environ Sustainability 2017. Available from: https://doi.org/10.47469/jees.2017.v04.100049.
[36] Rowe JW, Conner AH. Extractives in eastern hardwoods—a review. USDA For Serv Gen Tech Rep FPL; 1979.
[37] Guo XJ, Wang SR, Wang KG, Liu Q, Luo ZY. Influence of the extractives on mechanism of biomass pyrolysis. Ranliao Huaxue Xuebao/J Fuel Chem Technol 2010. Available from: https://doi.org/10.1016/s1872-5813(10)60019-9.
[38] Mészáros E, Jakab E, Várhegyi G. TG/MS, Py-GC/MS and THM-GC/MS study of the composition and thermal behavior of extractive components of *Robinia pseudoacacia*. J Anal Appl Pyrolysis 2007. Available from: https://doi.org/10.1016/j.jaap.2006.12.007.
[39] Kaur R, Gera P, Jha MK, Bhaskar T. Pyrolysis kinetics and thermodynamic parameters of castor (*Ricinus communis*) residue using thermogravimetric analysis. Bioresour Technol 2018. Available from: https://doi.org/10.1016/j.biortech.2017.11.077.
[40] Bridgwater A v, Carson P, Coulson M. A comparison of fast and slow pyrolysis liquids from mallee. Int J Glob Energy Issues 2007;27:204–16. Available from: https://doi.org/10.1504/IJGEI.2007.013655.
[41] Stamatov V, Honnery D, Soria J. Combustion properties of slow pyrolysis bio-oil produced from indigenous Australian species. Renew Energy 2006. Available from: https://doi.org/10.1016/j.renene.2005.10.004.

[42] Antal MJ, Grønli M. The art, science, and technology of charcoal production. Ind Eng Chem Res 2003. Available from: https://doi.org/10.1021/ie0207919.
[43] Prins MJ, Ptasinski KJ, Janssen FJJG. Torrefaction of wood. Part 1. Weight loss kinetics. J Anal Appl Pyrolysis 2006. Available from: https://doi.org/10.1016/j.jaap.2006.01.002.
[44] Bridgeman TG, Jones JM, Shield I, Williams PT. Torrefaction of reed canary grass, wheat straw and willow to enhance solid fuel qualities and combustion properties. Fuel 2008. Available from: https://doi.org/10.1016/j.fuel.2007.05.041.
[45] van der Stelt MJC, Gerhauser H, Kiel JHA, Ptasinski KJ. Biomass upgrading by torrefaction for the production of biofuels: a review. Biomass Bioenergy 2011. Available from: https://doi.org/10.1016/j.biombioe.2011.06.023.
[46] Oasmaa A, Kuoppala E, Elliott DC. Development of the basis for an analytical protocol for feeds and products of bio-oil hydrotreatment. Energy Fuels 2012. Available from: https://doi.org/10.1021/ef300252y.
[47] Hornung A. Intermediate pyrolysis of biomass. Woodhead Publishing Limited; 2013. Available from: https://doi.org/10.1533/9780857097439.2.172.
[48] Bridgwater AV. Renewable fuels and chemicals by thermal processing of biomass. Chem Eng J 2003. Available from: https://doi.org/10.1016/S1385-8947(02)00142-0.
[49] Kan T, Strezov V, Evans TJ. Lignocellulosic biomass pyrolysis: a review of product properties and effects of pyrolysis parameters. Renew Sustain Energy Rev 2016. Available from: https://doi.org/10.1016/j.rser.2015.12.185.
[50] Hosoya T, Kawamoto H, Saka S. Pyrolysis behaviors of wood and its constituent polymers at gasification temperature. J Anal Appl Pyrolysis 2007. Available from: https://doi.org/10.1016/j.jaap.2006.08.008.
[51] van de Velden M, Baeyens J, Brems A, Janssens B, Dewil R. Fundamentals, kinetics and endothermicity of the biomass pyrolysis reaction. Renew Energy 2010. Available from: https://doi.org/10.1016/j.renene.2009.04.019.
[52] Lin YC, Cho J, Tompsett GA, Westmoreland PR, Huber GW. Kinetics and mechanism of cellulose pyrolysis. J Phys Chem C 2009. Available from: https://doi.org/10.1021/jp906702p.
[53] Scheirs J, Camino G, Tumiatti W. Overview of water evolution during the thermal degradation of cellulose. Eur Polym J 2001. Available from: https://doi.org/10.1016/S0014-3057(00)00211-1.
[54] Worasuwannarak N, Sonobe T, Tanthapanichakoon W. Pyrolysis behaviors of rice straw, rice husk, and corncob by TG-MS technique. J Anal Appl Pyrolysis 2007. Available from: https://doi.org/10.1016/j.jaap.2006.08.002.
[55] Wang S, Ru B, Lin H, Luo Z. Degradation mechanism of monosaccharides and xylan under pyrolytic conditions with theoretic modeling on the energy profiles. Bioresour Technol 2013. Available from: https://doi.org/10.1016/j.biortech.2013.06.026.
[56] Shafizadeh F, McGinnis GD, Philpot CW. Thermal degradation of xylan and related model compounds. Carbohydr Res 1972. Available from: https://doi.org/10.1016/S0008-6215(00)82742-1.
[57] di Blasi C, Branca C, Galgano A. Biomass screening for the production of furfural via thermal decomposition. Ind Eng Chem Res 2010. Available from: https://doi.org/10.1021/ie901731u.
[58] Räisänen U, Pitkänen I, Halttunen H, Hurtta M. Formation of the main degradation compounds from arabinose, xylose, mannose and arabinitol during pyrolysis. J Therm Anal Calorim 2003. Available from: https://doi.org/10.1023/A:1024557011975.
[59] Patwardhan PR, Brown RC, Shanks BH. Product distribution from the fast pyrolysis of hemicellulose. ChemSusChem 2011. Available from: https://doi.org/10.1002/cssc.201000425.
[60] Couhert C, Commandre JM, Salvador S. Is it possible to predict gas yields of any biomass after rapid pyrolysis at high temperature from its composition in cellulose, hemicellulose and lignin? Fuel 2009. Available from: https://doi.org/10.1016/j.fuel.2008.09.019.
[61] Jensen A, Dam-Johansen K, Wójtowicz MA, Serio MA. TG-FTIR study of the influence of potassium chloride on wheat straw pyrolysis. Energy Fuels 1998. Available from: https://doi.org/10.1021/ef980008i.
[62] Kawamoto H. Lignin pyrolysis reactions. J Wood Sci 2017. Available from: https://doi.org/10.1007/s10086-016-1606-z.
[63] Patwardhan PR, Brown RC, Shanks BH. Understanding the fast pyrolysis of lignin. Chem Sus Chem 2011. Available from: https://doi.org/10.1002/cssc.201100133.

[64] Liu Q, Wang S, Zheng Y, Luo Z, Cen K. Mechanism study of wood lignin pyrolysis by using TG-FTIR analysis. J Anal Appl Pyrolysis 2008. Available from: https://doi.org/10.1016/j.jaap.2008.03.007.
[65] Basu P. Combustion and gasification in fluidized beds. Available from: <https://doi.org/10.1201/9781420005158>; 2006.
[66] Overview—Ensyn—renewable fuels and chemicals from non-food biomass. Available from: <http://www.ensyn.com/overview.html>; n.d. [accessed 10.01.21].
[67] Badger PC, Fransham P. Use of mobile fast pyrolysis plants to densify biomass and reduce biomass handling costs—a preliminary assessment. Biomass Bioenergy 2006. Available from: https://doi.org/10.1016/j.biombioe.2005.07.011.
[68] Garcia-Nunez JA, Pelaez-Samaniego MR, Garcia-Perez ME, Fonts I, Abrego J, Westerhof RJM, et al. Historical developments of pyrolysis reactors: a review. Energy Fuels 2017. Available from: https://doi.org/10.1021/acs.energyfuels.7b00641.
[69] Xiu S, Shahbazi A. Bio-oil production and upgrading research: a review. Renew Sustain Energy Rev 2012. Available from: https://doi.org/10.1016/j.rser.2012.04.028.
[70] Krutof A, Hawboldt KA. Upgrading of biomass sourced pyrolysis oil review: focus on co-pyrolysis and vapour upgrading during pyrolysis. Biomass Convers Biorefinery 2018. Available from: https://doi.org/10.1007/s13399-018-0326-6.
[71] Baldwin RM, Feik CJ. Bio-oil stabilization and upgrading by hot gas filtration. Energy Fuels 2013. Available from: https://doi.org/10.1021/ef400177t.
[72] Pawar A, Panwar NL, Salvi BL. Comprehensive review on pyrolytic oil production, upgrading and its utilization. J Mater Cycles Waste Manag 2020. Available from: https://doi.org/10.1007/s10163-020-01063-w.
[73] Shihadeh AL, Engineering SMM, Technology SM, Supervisor T. Rural electrification from local resources: biomass pyrolysis oil combustion in a direct injection diesel engine. Chem Eng 1999. Available from: https://dspace.mit.edu/handle/1721.1/43601.
[74] Chiaramonti D, Bonini M, Fratini E, Tondi G, Gartner K, Bridgwater AV, et al. Development of emulsions from biomass pyrolysis liquid and diesel and their use in engines—part 1: emulsion production. Biomass Bioenergy 2003. Available from: https://doi.org/10.1016/S0961-9534(02)00183-6.
[75] Chiaramonti D, Bonini M, Fratini E, Tondi G, Gartner K, Bridgwater AV, et al. Development of emulsions from biomass pyrolysis liquid and diesel and their use in engines—part 2: tests in diesel engines. Biomass Bioenergy 2003. Available from: https://doi.org/10.1016/S0961-9534(02)00184-8.
[76] Ikura M, Stanciulescu M, Hogan E. Emulsification of pyrolysis derived bio-oil in diesel fuel. Biomass Bioenergy 2003. Available from: https://doi.org/10.1016/S0961-9534(02)00131-9.
[77] Xiaoxiang J, Ellis N. Upgrading bio-oil through emulsification with biodiesel: mixture production. Energy Fuels 2010. Available from: https://doi.org/10.1021/ef9010669.
[78] Mirmiran S, Sawatzky H, Toomer AED. United States patent (19); 1998.
[79] Diebold JP, Czernik S. Additives to lower and stabilize the viscosity of pyrolysis oils during storage. Energy Fuels 1997. Available from: https://doi.org/10.1021/ef9700339.
[80] Bridgwater A. Fast pyrolysis of biomass for the production of liquids. Biomass Combust Sci Technol Eng 2013. Available from: https://doi.org/10.1533/9780857097439.2.130.
[81] Huber GW, Dumesic JA. An overview of aqueous-phase catalytic processes for production of hydrogen and alkanes in a biorefinery. Catal Today 2006. Available from: https://doi.org/10.1016/j.cattod.2005.10.010.
[82] Fisk C, Crofcheck C, Crocker M, Andrews R, Storey J, Lewis Sr S. Novel approaches to catalytic upgrading of bio-oil; 2013. Available from: <https://doi.org/10.13031/2013.21981>.
[83] Liao HT, Ye XN, Lu Q, Dong CQ. Overview of bio-oil upgrading via catalytic cracking. Adv Mater Res 2014;827:25–9. Available from: https://doi.org/10.4028/http://www.scientific.net/AMR.827.25.
[84] Lee AF, Bennett JA, Manayil JC, Wilson K. Heterogeneous catalysis for sustainable biodiesel production via esterification and transesterification. Chem Soc Rev 2014. Available from: https://doi.org/10.1039/c4cs00189c.
[85] Wu L, Hu X, Wang S, Mahmudul Hasan MD, Jiang S, Li T, et al. Acid-treatment of bio-oil in methanol: the distinct catalytic behaviours of a mineral acid catalyst and a solid acid catalyst. Fuel 2018;212:412–21. Available from: https://doi.org/10.1016/j.fuel.2017.10.049.

[86] Liu Y, Li Z, Leahy JJ, Kwapinski W. Catalytically upgrading bio-oil via esterification. Energy Fuels 2015;29:3691–8. Available from: https://doi.org/10.1021/acs.energyfuels.5b00163.

[87] Zhang P, Sun Y, Zhang Q, Guo Y, Song D. Upgrading of pyrolysis biofuel via esterification of acetic acid with benzyl alcohol catalyzed by BrØnsted acidic ionic liquid functionalized ethyl-bridged organosilica hollow nanospheres. Fuel 2018;228:175–86. Available from: https://doi.org/10.1016/j.fuel.2018.04.107.

[88] Manayil JC, Inocencio CVM, Lee AF, Wilson K. Mesoporous sulfonic acid silicas for pyrolysis bio-oil upgrading via acetic acid esterification. Green Chem 2016;18:1387–94. Available from: https://doi.org/10.1039/c5gc01889g.

[89] Lee Y, Shafaghat H, Kim Jk, Jeon JK, Jung SC, Lee IG, et al. Upgrading of pyrolysis bio-oil using WO_3/ZrO_2 and Amberlyst catalysts: evaluation of acid number and viscosity. Korean J Chem Eng 2017;34:2180–7. Available from: https://doi.org/10.1007/s11814-017-0126-x.

[90] Wang JJ, Chang J, Fan J. Experimental study on catalytic esterification of bio-oil by ion exchange resins. Ranliao Huaxue Xuebao/J Fuel Chem Technol 2010;38:560–4. Available from: https://doi.org/10.1016/S1872-5813(10)60045-X.

[91] Peng J, Chen P, Lou H, Zheng X. Catalytic upgrading of bio-oil by HZSM-5 in sub- and super-critical ethanol. Bioresour Technol 2009;100:3415–18. Available from: https://doi.org/10.1016/j.biortech.2009.02.007.

[92] Xu J, Jiang J, Dai W, Zhang T, Xu Y. Bio-oil upgrading by means of ozone oxidation and esterification to remove water and to improve fuel characteristics. Energy Fuels 2011;25:1798–801. Available from: https://doi.org/10.1021/ef101726g.

[93] Saber M, Nakhshiniev B, Yoshikawa K. A review of production and upgrading of algal bio-oil. Renew Sustain Energy Rev 2016;58:918–30. Available from: https://doi.org/10.1016/j.rser.2015.12.342.

[94] Zhang Q, Chang J, Wang TJ, Xu Y. Upgrading bio-oil over different solid catalysts. Energy Fuels 2006;20:2717–20. Available from: https://doi.org/10.1021/ef060224o.

[95] Gandarias I, Barrio VL, Requies J, Arias PL, Cambra JF, Güemez MB. From biomass to fuels: hydrotreating of oxygenated compounds. Int J Hydrogen Energy 2008;33:3485–8. Available from: https://doi.org/10.1016/j.ijhydene.2007.12.070.

[96] Melero JA, Clavero MM, Calleja G, García A, Miravalles R, Galindo T. Production of biofuels via the catalytic cracking of mixtures of crude vegetable oils and nonedible animal fats with vacuum gas oil. Energy Fuels 2010;24:707–17. Available from: https://doi.org/10.1021/ef900914e.

[97] Adjaye JD, Bakhshi NN. Catalytic conversion of a biomass-derived oil to fuels and chemicals I: model compound studies and reaction pathways. Biomass Bioenergy 1995;8:131–49. Available from: https://doi.org/10.1016/0961-9534(95)00018-3.

[98] de Miguel Mercader F, Groeneveld MJ, Kersten SRA, Way NWJ, Schaverien CJ, Hogendoorn JA. Production of advanced biofuels: co-processing of upgraded pyrolysis oil in standard refinery units. Appl Catal B Environ 2010;96:57–66. Available from: https://doi.org/10.1016/j.apcatb.2010.01.033.

[99] Wang Y, Fang Y, He T, Hu H, Wu J. Hydrodeoxygenation of dibenzofuran over noble metal supported on mesoporous zeolite. Catal Commun 2011;12:1201–5. Available from: https://doi.org/10.1016/j.catcom.2011.04.010.

[100] Hong DY, Miller SJ, Agrawal PK, Jones CW. Hydrodeoxygenation and coupling of aqueous phenolics over bifunctional zeolite-supported metal catalysts. Chem Commun 2010;46:1038–40. Available from: https://doi.org/10.1039/b918209h.

[101] Runnebaum RC, Lobo-Lapidus RJ, Nimmanwudipong T, Block DE, Gates BC. Conversion of anisole catalyzed by platinum supported on alumina: the reaction network. Energy Fuels 2011;25:4776–85. Available from: https://doi.org/10.1021/ef2010699.

[102] Cheng S, Wei L, Julson J, Rabnawaz M. Upgrading pyrolysis bio-oil through hydrodeoxygenation (HDO) using non-sulfided Fe-Co/SiO_2 catalyst. Energy Convers Manag 2017;150:331–42. Available from: https://doi.org/10.1016/j.enconman.2017.08.024.

[103] Zhang S, Yan Y, Li T, Ren Z. Upgrading of liquid fuel from the pyrolysis of biomass. Bioresour Technol 2005;96:545–50. Available from: https://doi.org/10.1016/j.biortech.2004.06.015.

[104] Fisk CA, Morgan T, Ji Y, Crocker M, Crofcheck C, Lewis SA. Bio-oil upgrading over platinum catalysts using in situ generated hydrogen. Appl Catal A Gen 2009;358:150–6. Available from: https://doi.org/10.1016/j.apcata.2009.02.006.
[105] Romero Y, Richard F, Brunet S. Hydrodeoxygenation of 2-ethylphenol as a model compound of bio-crude over sulfided Mo-based catalysts: promoting effect and reaction mechanism. Appl Catal B Environ 2010;98:213–23. Available from: https://doi.org/10.1016/j.apcatb.2010.05.031.
[106] Corma A, Huber GW, Sauvanaud L, O'Connor P. Processing biomass-derived oxygenates in the oil refinery: catalytic cracking (FCC) reaction pathways and role of catalyst. J Catal 2007;247:307–27. Available from: https://doi.org/10.1016/j.jcat.2007.01.023.
[107] Twaiq FA, Zabidi NAM, Mohamed AR, Bhatia S. Catalytic conversion of palm oil over mesoporous aluminosilicate MCM-41 for the production of liquid hydrocarbon fuels. Fuel Process Technol 2003;84:105–20. Available from: https://doi.org/10.1016/S0378-3820(03)00048-1.
[108] Adjaye JD, Bakhshi NN. Production of hydrocarbons by catalytic upgrading of a fast pyrolysis bio-oil. Part II: comparative catalyst performance and reaction pathways. Fuel Process Technol 1995;45:185–202. Available from: https://doi.org/10.1016/0378-3820(95)00040-E.
[109] Adam J, Antonakou E, Lappas A, Stöcker M, Nilsen MH, Bouzga A, et al. In situ catalytic upgrading of biomass derived fast pyrolysis vapours in a fixed bed reactor using mesoporous materials. Microporous Mesoporous Mater 2006. Available from: https://doi.org/10.1016/j.micromeso.2006.06.021.
[110] Shakya R, Adhikari S, Mahadevan R, Hassan EB, Dempster TA. Catalytic upgrading of bio-oil produced from hydrothermal liquefaction of *Nannochloropsis* sp. Bioresour Technol 2018;252:28–36. Available from: https://doi.org/10.1016/j.biortech.2017.12.067.
[111] Koike N, Hosokai S, Takagaki A, Nishimura S, Kikuchi R, Ebitani K, et al. Upgrading of pyrolysis bio-oil using nickel phosphide catalysts. J Catal 2016;333:115–26. Available from: https://doi.org/10.1016/j.jcat.2015.10.022.
[112] Wang H, Male J, Wang Y. Recent advances in hydrotreating of pyrolysis bio-oil and its oxygen-containing model compounds. ACS Catal 2013;3:1047–70. Available from: https://doi.org/10.1021/cs400069z.
[113] Zacher AH, Olarte Mv, Santosa DM, Elliott DC, Jones SB. A review and perspective of recent bio-oil hydrotreating research. Green Chem 2014;16:491–515. Available from: https://doi.org/10.1039/c3gc41382a.
[114] Wang Y, He T, Liu K, Wu J, Fang Y. From biomass to advanced bio-fuel by catalytic pyrolysis/hydro-processing: hydrodeoxygenation of bio-oil derived from biomass catalytic pyrolysis. Bioresour Technol 2012;108:280–4. Available from: https://doi.org/10.1016/j.biortech.2011.12.132.
[115] Tanneru SK, Steele PH. Pretreating bio-oil to increase yield and reduce char during hydrodeoxygenation to produce hydrocarbons. Fuel 2014;133:326–31. Available from: https://doi.org/10.1016/j.fuel.2014.05.026.
[116] Li N, Tompsett GA, Huber GW. Renewable high-octane gasoline by aqueous-phase hydrodeoxygenation of C5 and C6 carbohydrates over Pt/zirconium phosphate catalysts. ChemSusChem 2010;3:1154–7. Available from: https://doi.org/10.1002/cssc.201000140.
[117] Dickerson T, Soria J. Catalytic fast pyrolysis: a review. Energies 2013;6:514–38. Available from: https://doi.org/10.3390/en6010514.
[118] Wang H, Male J, Wang Y. Recent advances in hydrotreating of pyrolysis bio-oil and its oxygen-containing model compounds. ACS Catal 2013;3:1047–70. Available from: https://doi.org/10.1021/cs400069z.
[119] Sanna A, Vispute TP, Huber GW. Hydrodeoxygenation of the aqueous fraction of bio-oil with Ru/C and Pt/C catalysts. Appl Catal B Environ 2015;165:446–56. Available from: https://doi.org/10.1016/j.apcatb.2014.10.013.
[120] Baker EG, Elliott DC. Catalytic upgrading of biomass pyrolysis oils. Res Thermo Bio Conv 1988.
[121] French RJ, Stunkel J, Baldwin RM. Mild hydrotreating of bio-oil: effect of reaction severity and fate of oxygenated species. Energy Fuels 2011;25:3266–74. Available from: https://doi.org/10.1021/ef200462v.
[122] Zhang L, Liu R, Yin R, Mei Y. Upgrading of bio-oil from biomass fast pyrolysis in China: a review. Renew Sustain Energy Rev 2013;24:66–72. Available from: https://doi.org/10.1016/j.rser.2013.03.027.

[123] Tang Y, Miao S, Mo L, Zheng X, Shanks BH. One-step hydrogenation/esterification activity enhancement over bifunctional mesoporous organic-inorganic hybrid silicas. Top Catal 2013;56:1804–13. Available from: https://doi.org/10.1007/s11244-013-0117-z.

[124] Tang Y, Miao S, Shanks BH, Zheng X. Bifunctional mesoporous organic-inorganic hybrid silica for combined one-step hydrogenation/esterification. Appl Catal A Gen 2010;375:310–17. Available from: https://doi.org/10.1016/j.apcata.2010.01.015.

[125] Yaman S. Pyrolysis of biomass to produce fuels and chemical feedstocks. Energy Convers Manag 2004. Available from: https://doi.org/10.1016/S0196-8904(03)00177-8.

[126] Paul RC. Methane steam reforming. Catal A Z 2020. Available from: https://doi.org/10.1002/9783527809080.cataz10591.

[127] Vagia EC, Lemonidou AA. Hydrogen production via steam reforming of bio-oil components over calcium aluminate supported nickel and noble metal catalysts. Appl Catal A Gen 2008;351:111–21. Available from: https://doi.org/10.1016/j.apcata.2008.09.007.

[128] Boskovic G, Baerns M. Catalyst deactivation. Springer Ser Chem Phys 2004;75:477–503. Available from: https://doi.org/10.1007/978-3-662-05981-4_14.

[129] Román Galdámez J, García L, Bilbao R. Hydrogen production by steam reforming of bio-oil using coprecipitated Ni-Al catalysts. Acetic acid as a model compound. Energy Fuels 2005;19:1133–42. Available from: https://doi.org/10.1021/ef049718g.

[130] Xu Q, Lan P, Zhang B, Ren Z, Yan Y. Hydrogen production via catalytic steam reforming of fast pyrolysis bio-oil in a fluidized-bed reactor. Energy Fuels 2010;24:6456–62. Available from: https://doi.org/10.1021/ef1010995.

[131] Yan CF, Cheng FF, Hu RR. Hydrogen production from catalytic steam reforming of bio-oil aqueous fraction over Ni/CeO_2-ZrO_2 catalysts. Int J Hydrogen Energy 2010;35:11693–9. Available from: https://doi.org/10.1016/j.ijhydene.2010.08.083.

[132] Marquevich M, Czernik S, Chornet E, Montané D. Hydrogen from biomass: Steam reforming of model compounds of fast-pyrolysis oil. Energy Fuels 1999;13:1160–6. Available from: https://doi.org/10.1021/ef990034w.

[133] Basagiannis AC, Verykios XE. Reforming reactions of acetic acid on nickel catalysts over a wide temperature range. Appl Catal A Gen 2006;308:182–93. Available from: https://doi.org/10.1016/j.apcata.2006.04.024.

[134] Hu X, Lu G. Investigation of steam reforming of acetic acid to hydrogen over Ni-Co metal catalyst. J Mol Catal A Chem 2007;261:43–8. Available from: https://doi.org/10.1016/j.molcata.2006.07.066.

[135] Yan CF, Hu EY, Cai CL. Hydrogen production from bio-oil aqueous fraction with in situ carbon dioxide capture. Int J Hydrogen Energy 2010;35:2612–16. Available from: https://doi.org/10.1016/j.ijhydene.2009.04.016.

[136] Takanabe K, Aika KI, Seshan K, Lefferts L. Sustainable hydrogen from bio-oil – Steam reforming of acetic acid as a model oxygenate. J Catal 2004;227:101–8. Available from: https://doi.org/10.1016/j.jcat.2004.07.002.

[137] Yamazaki T, Matsuki K. Steam reforming reaction of biomass-derived substances over nanoporous Ru/ZrO_2 catalysts (Part 1) steam reforming reaction of acetic acid. J Jpn Pet Inst 2006;49:246–55. Available from: https://doi.org/10.1627/jpi.49.246.

[138] Rioche C, Kulkarni S, Meunier FC, Breen JP, Burch R. Steam reforming of model compounds and fast pyrolysis bio-oil on supported noble metal catalysts. Appl Catal B Environ 2005;61:130–9. Available from: https://doi.org/10.1016/j.apcatb.2005.04.015.

[139] Xu C, Etcheverry T. Hydro-liquefaction of woody biomass in sub- and super-critical ethanol with iron-based catalysts. Fuel 2008;87:335–45. Available from: https://doi.org/10.1016/j.fuel.2007.05.013.

[140] Minami E, Saka S. Decomposition behavior of woody biomass in water-added supercritical methanol. J Wood Sci 2005;51:395–400. Available from: https://doi.org/10.1007/s10086-004-0670-y.

[141] Minami E, Saka S. Comparison of the decomposition behaviors of hardwood and softwood in supercritical methanol. J Wood Sci 2003;49:73–8. Available from: https://doi.org/10.1007/s100860300012.

[142] Xiu S, Shahbazi A, Wang L, Wallace CW. Supercritical ethanol liquefaction of swine manure for bio-oils production. Am J Eng Appl Sci 2010;3:494–500.

[143] Liu Z, Zhang FS. Effects of various solvents on the liquefaction of biomass to produce fuels and chemical feedstocks. Energy Convers Manag 2008;49:3498–504. Available from: https://doi.org/10.1016/j.enconman.2008.08.009.
[144] Heitz M, Brown A, Chornet E. Solvent effects on liquefaction: solubilization profiles of a Canadian prototype wood, populus deltoides, in the presence of different solvents. Can J Chem Eng 1994;72:1021–7. Available from: https://doi.org/10.1002/cjce.5450720612.
[145] Omar S, Yang Y, Wang J. A review on catalytic & non-catalytic bio-oil upgrading in supercritical fluids. Front Chem Sci Eng 2021;15:4–17. Available from: https://doi.org/10.1007/s11705-020-1933-x.
[146] Peng J, Chen P, Lou H, Zheng X. Upgrading of bio-oil over aluminum silicate in supercritical ethanol. Energy Fuels 2008;22:3489–92. Available from: https://doi.org/10.1021/ef8001789.
[147] Dang Q, Luo Z, Zhang J, Wang J, Chen W, Yang Y. Experimental study on bio-oil upgrading over $Pt/SO_4^{2-}/ZrO_2$/SBA-15 catalyst in supercritical ethanol. Fuel 2013;103:683–92. Available from: https://doi.org/10.1016/j.fuel.2012.06.082.
[148] di Sanzo G, Mehariya S, Martino M, Larocca V, Casella P, Chianese S, et al. Supercritical carbon dioxide extraction of astaxanthin, lutein, and fatty acids from *Haematococcus pluvialis* microalgae. Mar Drugs 2018;16. Available from: https://doi.org/10.3390/md16090334.
[149] Molino A, Iovine A, Casella P, Mehariya S, Chianese S, Cerbone A, et al. Microalgae characterization for consolidated and new application in human food, animal feed and nutraceuticals. Int J Environ Res Public Health 2018;15:1–21. Available from: https://doi.org/10.3390/ijerph15112436.
[150] Poojary MM, Barba FJ, Aliakbarian B, Donsì F, Pataro G, Dias DA, et al. Innovative alternative technologies to extract carotenoids from microalgae and seaweeds. Mar Drugs 2016;14:1–34. Available from: https://doi.org/10.3390/md14110214.
[151] Pereira CG, Meireles MAA. Supercritical fluid extraction of bioactive compounds: fundamentals, applications and economic perspectives. Food Bioprocess Technol 2010;3:340–72. Available from: https://doi.org/10.1007/s11947-009-0263-2.
[152] Ruen-Ngam D, Shotipruk A, Pavasant P, Machmudah S, Goto M. Selective extraction of lutein from alcohol treated *Chlorella vulgaris* by supercritical CO_2. Chem Eng Technol 2012;35:255–60. Available from: https://doi.org/10.1002/ceat.201100251.
[153] Bari S. Performance, combustion and emission tests of a metro-bus running on biodiesel-ULSD blended (B20) fuel. Appl Energy 2014;124:35–43. Available from: https://doi.org/10.1016/j.apenergy.2014.03.007.
[154] Solantausta Y, Nylund NO, Westerholm M, Koljonen T, Oasmaa A. Wood-pyrolysis oil as fuel in a diesel-power plant. Bioresour Technol 1993;46:177–88. Available from: https://doi.org/10.1016/0960-8524(93)90071-I.
[155] Spath PL, Mann MK. Biomass power and conventional fossil systems with and without CO_2 sequestration—comparing the energy balance, greenhouse gas emissions and economics. Contract 2004;30:38.
[156] Chiaramonti D, Oasmaa A, Solantausta Y. Power generation using fast pyrolysis liquids from biomass. Renew Sustain Energy Rev 2007;11:1056–86. Available from: https://doi.org/10.1016/j.rser.2005.07.008.
[157] Heat—BTG bioliquids, Available from: <https://www.btg-bioliquids.com/market/heat/>; n.d. [accessed 15.04.21].

CHAPTER 2

Biomass pyrolysis system based on life cycle assessment and Aspen plus analysis and kinetic modeling

2.1 Introduction

Fossil fuels are the most predominantly used primary energy source globally until today. Oil, natural gas, and coal are the most widely used forms of energy among the various types of fossil fuel energy sources. However, the usage of fossil fuels has several demerits, including the evolution of noxious hydrocarbon gases and contributing to global warming. Moreover, the limited availability indicates the necessity for us to move towards sustainable energy sources. The climate panel at the United Nations Convention has announced an ambitious goal of reducing the level of greenhouse gas (GHG) emissions by 50%–80% by the year 2050 [1]. Hence, the future of the energy sector relies on investigating novel materials and energy sources to supplant fossil fuels and move towards the era of renewable energy [2,3].

Among the different types of renewable energy sources, extracting energy from biomass is one of the most predominantly used methods for energy generation. Various treatment methods, including thermochemical, biological, and mechanical processes, convert the hydrocarbons present in the biomass into value-added substances and fuels [4,5]. Compared to the other two approaches, thermochemical energy conversion has higher efficiency and low costs and thus a promising methodology for this purpose [6].

The biological wastes that are constituted only of organic material can be considered for thermochemical energy conversion. However, a large quantity of trash is generated in the form of plastics or textiles (municipal solid wastes—MSWs), which are being discarded in large quantities in landfills. These wastes, which are relatively low in cost, can be converted into useful energy forms through copyrolysis with biomass. Several feedstocks are being tested with MSWs to extract energy through the process of pyrolysis and have the potential to combat the global energy challenge. However, these energy generation processes are not yet implemented on a large scale due to the existing optimized petroleum-based processes and the economic uncertainties. Moreover, this waste biomass energy conversion process requires enormous capital investments due to the complexity of many unit operations. It is also laborious compared to the existing methods.

A Thermo-Economic Approach to Energy From Waste
DOI: https://doi.org/10.1016/B978-0-12-824357-2.00006-1

Therefore a detailed study and investigation of the process intensification are required to overcome these challenges. While there are numerous experimental studies on pyrolysis and copyrolysis of biomass, there is not much work done on mathematical/process modeling. Mathematical modeling tools can simulate the real-time conditions of the entire process and validate the influence of the operating parameters on the performance of the process. However, modeling these processes can prove highly challenging as it involves considering the chemical, thermodynamic, and fluid dynamic properties during the process.

Furthermore, there is an urgency to assess large-scale units' energy generation potential for the successful commercial implementation of pyrolysis technology. Since there is a global concern on climate change due to GHG emissions, the assessment of footprints of CO_2 should also be done alongside the energy analysis. For the avoidance of unintended detrimental environmental consequences of a new technology accounting for the environmental impact of the process should be done. The most appropriate methodology for this purpose is life cycle analysis (LCA). Similarly, the energy output-input ratio showing the net energy return from a new process is also necessary before its actual implementation. Thus an analysis accounting for both environmental impact and energy return, called energy environment analysis, following the basic principles of LCA, is helpful to serve both purposes.

Similarly, Aspen Plus software has been used to simulate both lab-scale and large-scale pyrolyzers. The prediction of both Aspen Plus models has been compared with the small reactor's experimental data, and the valid model is chosen for further analysis of a plant. It is also intended to carry out simultaneously the energy and CO_2 emission analyses of the large pyrolysis setup using the basic principles of LCA. Also, presently used techniques for the kinetic modeling of the pyrolysis process will be explained in detail in the upcoming sections. Also, the last part of this chapter discussed applications of biomass pyrolysis products, such as bio-oil and biochar, are presented.

2.2 Current Indian scenario of waste-to-energy conversion technologies

The rapid rise in the global population has led to an unprecedented increase in demand for energy sources. An unintended consequence of this rise is the generation of waste products, most notably MSWs. Various types of products including garbage, construction and demolition waste, nonhazardous industrial waste, biochemical wastes, and sanitary residue [7,8]. The majority of the MSW comes from improper recycling and segregation of household wastes leading to increased landfill disposal. Furthermore, besides being a dumping ground for inefficient MSW, landfills have become a dumping ground for recyclable wastes. A 2012 investigation by the World Bank found a waste generation to the tune of about 1.5 billion tonnes in a calendar year. This generation is likely to rise by 0.9 billion tonnes by 2025.

High per capita income indicates greater buying power for the citizens and, consequently, higher waste generation. Hence, more prosperous nations contribute to greater amounts of waste than poorer ones. Even in developing nations, such as India, rapid urbanization has contributed to waste generation. It is estimated that more than 30% of our country's population lives in cities. Furthermore, metropolitan cities, such as Delhi, Kolkata, and Mumbai, have a population greater than 0.1% of the country's population. The population growth is not limited to these areas. Fifty-three cities have more than 1 million population, and more than 400 urban areas have a population of 500,000 or more (Census 2011). Thus this growth in the urban population has resulted in a significant rise in solid waste in urban areas. Steady population rise contributes to 1.3% annual rise in the waste generation [7,9]. Table 2.1 shows the waste generation and collection data for various states in India.

In the postindustrial era, the first attempts at converting waste products into energy sources were made. The gaseous product produced from thermochemical conversion of solid waste products was termed as syngas. Initial applications of this product were in urban lighting infrastructure. For this reason, the gas was popularly known as "city gas." The gasometers that stored this city's gas can be found in large cities even to this day. The development of sustainable cities is essential for efficient waste management

Table 2.1 Collection and production data of municipal solid waste during the year 2017–18 [18].

S. no.	States	Urban local bodies	Solid waste-generated (Ton per day)	Solid waste-collected (Ton per day)
1	Arunachal Pradesh	17 urban divisions 2 municipal councils	203.96	166.82
2	Madhya Pradesh	378	7212	6537
3	Meghalaya	7	210	175
4	Nagaland	32	348	252
5	Orissa	114	539.44	471.58
6	Sikkim	7	73.34	63
7	Tirpura	20		379.2
8	Jammu & Kashmir	Jammu division: 78 Kashmir division: 42	666.68 930.6	617.83 833.3
9	Chattisgarh	166		2000
10	Uttarakhand	101	1099	1099
11	Chandigarh	1	500	463
12	Tamil Nadu	12 corporations 124 municipalities 528 town panchayats	15,176.612	14,568.333
13	Telangana	73	7804	7023

[10,11]. However, harvesting energy from MSW might be even more beneficial to the alternative energy industry [10–13]. Although the large amounts of MSW in urban areas have the potential to meet cities energy demand, their use is being limited due to our over-reliance on fossil fuels.

Although fossil fuel-derived resources can meet the energy demands, their availability is limited. Furthermore, the processing costs associated with these resources cause a tremendous rise in their prices. Since the world energy demand is increasing due to technological progress, meeting the price of fossil fuel-derived commodities has become difficult [14,15]. Another reason for increasing the market share of renewable resources is the harmful side effects of fossil fuels. Usage of fossil fuels is directly correlated with an increase in GHG emissions. Thus the efficient use of MSW can reduce the impacts of global warming.

Furthermore, the land necessary for municipal waste disposal also reduces drastically [14–16]. Studies have shown that MSW can provide sufficient energy to sustain small communities and even industries [17]. The following sections will detail the several efforts made by researchers to use MSW for energy generation sustainably.

2.3 From biomass to biofuel through pyrolysis

One of the first energy sources used by civilized humans was biomass. During the Iron Age, charcoal (a fuel produced from burning biomass) was used to melt ore to make iron. In a few developing nations, biomass is still being used to meet household energy demands. However, in the late 1970s, western countries began interested in energy production from biomass sources. There are three main pathways for energy recovery from biomass. These are combustion, gasification, and pyrolysis [19]. Among these processes, the only pyrolysis has the characteristic of being an endothermic process. The scope for development was, particularly in pyrolysis—earlier pyrolysis technologies suffered from the disadvantages of slow production, air pollution, and inefficient yield. Thus advancement in pyrolysis technology is necessary to make a profitable investment. Combustion refers to fuel oxidation, where the burning of the input products gives rise to heat energy. However, this process is highly inefficient and is a significant source of pollution [20,21].

On the other hand, gasification is a process of partial oxidation wherein gas products are formed from solid substrates. On the other hand, pyrolysis a combination of both gasification and combustion [21,22]. Therefore the pyrolysis process is both an independent and a hybrid thermochemical approach [23], wherein the solid substrate breaks down to give rise to liquid and gaseous products. During pyrolysis, the organic matter is energized in an inert environment, wherein complex series of simultaneous and sequential reactions occur. In pyrolysis, the thermal decomposition of the substrate begins at around 350º C and can go up to 800º C in the absence of air [24]. At the

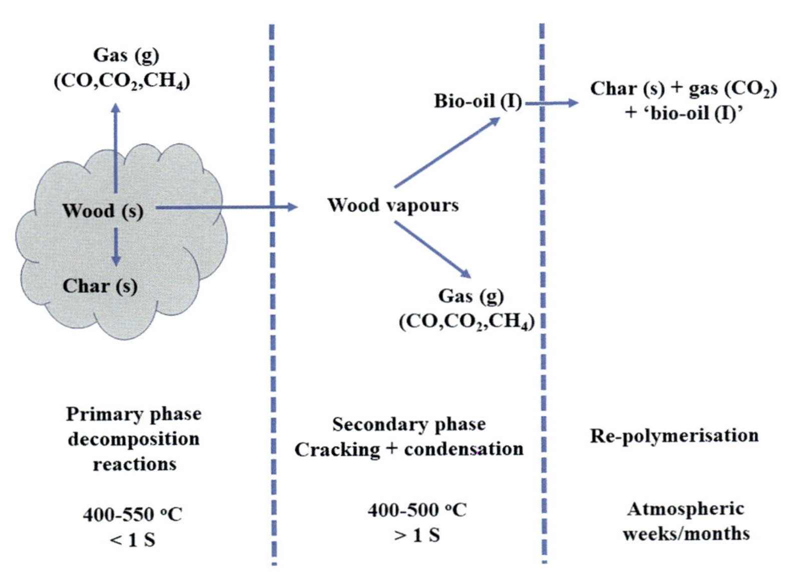

Figure 2.1 Woody biomass pyrolysis reaction schemes [27].

micro-scale, the links connecting the elements C, H, and O disintegrate during pyrolysis to produce gaseous, liquid (tars and oils), and solid products. The extent of decomposition and rate of reaction for pyrolysis depends on several factors, including heating rate, pressure, temperature, and reactor environment.

The reaction pathways for woody biomass pyrolysis are shown in Fig. 2.1. These pathways are in the form of lumped reactions beginning with a first-order reaction. One study concluded that the volatile release rate at the initial stages of the pyrolysis process is nearly 10 times the rate of release in the next step [25]. However, despite the description provided above, pyrolysis cannot be described by one all-encompassing mechanism because different compounds have different substrate compositions [26].

2.4 Life cycle assessment methodology for pyrolysis-based bio-oil production

Life cycle assessment (LCA) is emerging as a popular method to estimate GHG emissions and other environmental factors during the formation of the product. Thus throughout the process, LCA can provide a qualitative and quantitative estimate of various impact parameters. In general, the process of conversion of woody biomass into desirable compounds includes the following stages: (1) biomass cultivation and transportation to plant site, (2) operation of onsite equipment and product upgradation, and (3) plant demolition and recycling (see Fig. 2.2). Due to the abundance of data, the first two phases

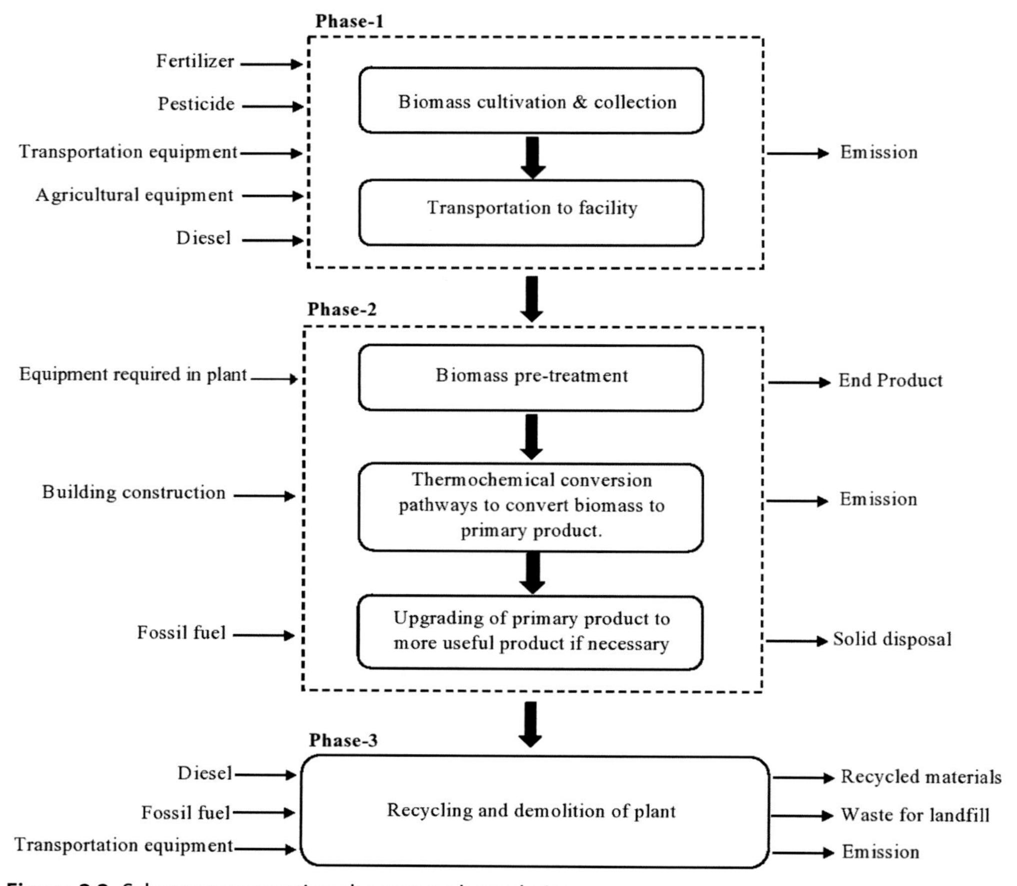

Figure 2.2 Scheme representing the system boundaries.

have been studied extensively. Past research on LCA was predominantly in technologies, such as pyrolysis, gasification, cofiring, and combustion. Among these technologies, pyrolysis has generated the most interest within LCA studies. However, there is a substantial gap in LCA research in carbonization and liquefaction. Various researchers employ various software, such as GH Genius, GREET (Greenhouse gases, Regulated Emissions, and Energy use in Transportation), Sigma Pro, and TEAM (Tools for Environmental Analysis and Management to collect inventory data) [28–30]. CML and Eco Indicator 95/99 are used to evaluate the environmental impact factors for various substrates for different system boundary conditions [31,32]. Among the several factors that impact LCA, the most important ones are the type of feedstock, the functional unit, the system boundary, and the environmental impact. Among the feedstocks, lignocellulosic biomass has found applications in LCA systems. In general, woody biomass consists of lignin, hemicellulose, and cellulose. However, these unit groups concentrations vary for different types of feedstocks. A higher fraction of cellulose for the pyrolysis process and a lower quantity of lignin are beneficial. Apart from the chemical composition, the collection and cultivation methods also influence the environmental impact. Within the same framework of cultivation methods, a new type of crop termed an "energy crop" was developed to yield an economical and sustainable energy source. These crops are also called short rotation or high yield crops genetically modified to compete as energy resources alongside crude oil-derived resources [33]. One of the significant disadvantages of these energy crops is the necessary employment of pesticides, giving rise to resistant insects, thereby harming crops that are not genetically treated [34]. Forest residues are another category of feedstocks being employed in LCA studies. The sources for these feedstocks include deadwood waste, waste from lumber mills, and extraction of timber from forests (postextraction). Forest residues share a few commonalities with wood residue, that is, they have similar heating values and moisture content. However, their ash content is much lower than wood residues.

Furthermore, the moisture content in forests varies depending on the season and climate [34]. The last type of biomass feedstocks is agricultural residues, which are leftover components of subsistence crops. Corn stover, wheat straw, and rice husk are some of the most popular agricultural residues with very low water content and very high heating value. The demand for these sources is quite high because of competing resources (animal fodder) and crop seasonality. It is also fairly common to use these residues to improve groundwater levels and soil fertility by shielding the soil from wind, rain, and sunlight [35]. Thus it is necessary to take these complex conditions into account while doing LCA.

2.4.1 Steps followed for studying LCA

LCA is a technique that analyses the impacts on the environment because of a process [35]. Its applications include the following:

- assessing the contribution of individual processes and steps in the total environmental impact factors, thereby improving the sustainability of these steps
- Comparing among various substrates for usage in thermochemical processes.

An LCA study comprises four steps, as shown in Fig. 2.3.

Stage 1: The purpose of defining a goal and a scope is to establish the boundaries of the LCA. The specifications about specific times and system comparisons are formulated here.

Stage 2: The collection of inventory data provides an account of the energy and material flows and their interaction with the environment and raw substrates. The subsidiary flow processes and other important processes are not described here.

Stage 3: The specifications provided during the inventory analysis are utilized in this stage. The outputs produced from the categories feed in the inventory analysis are explained here. The various impact parameters are also weighted and normalized.

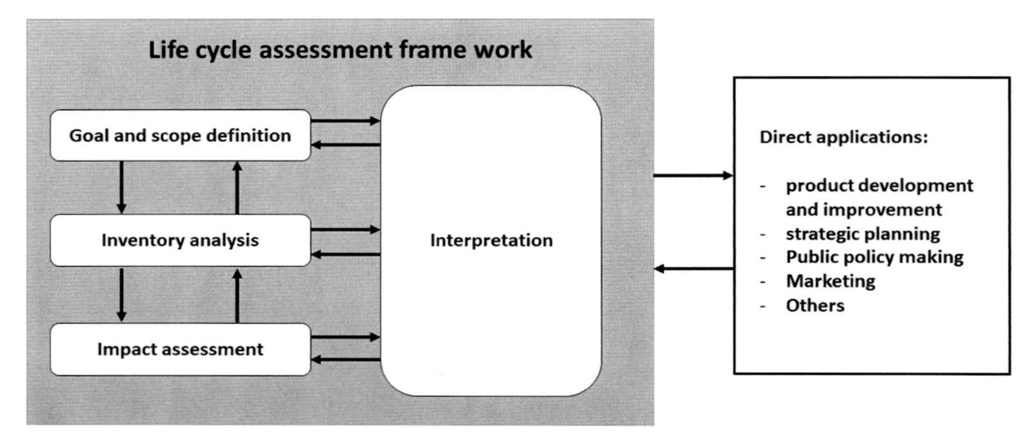

Figure 2.3 EN ISO 14040-mandated life cycle assessment stages.

Stage 4: This last step involves critical examination, result presentation, and data reliability and sensitivity evaluation.

Fig. 2.3 shows the ISO 14040 guidelines mandated stages.

When an LCA study is being conducted, the subsequent points need to be addressed:

The environmental impact of a thermochemical process can be evaluated by noting down the raw materials and input energy at every step and the resultant wastes and pollutants. The study's parameters are then analyzed for their influence on renewable and nonrenewable resource sustainability and biodiversity. This step is necessary to find out the environmental impact of the output parameters.

The employment of LCA can aid in the following steps:

- looking for the most amount of life cycles, for instance, the ones with the least environmental impact;
- recommending plans of action for NGOs and public and private organizations. Furthermore, one can determine the direction in strategic planning and product design;
- analyzing the environmental impact indicators of an organization and suggest steps to mitigate the impact; and
- eco-labeling formulation.

2.4.2 Setting require for LCA

Phases involved. The thermochemical conversion of woody biomass into by-products has three different phases concerning LCA. Most studies in the literature have employed system boundaries associated with phase 1 and phase 2. However, few researchers have performed complete analysis, that is, the "cradle-to-grave" studies. The first phase can be subdivided into two sections: the collection and cultivation of biomass and the transportation to the plant site. In this phase, the parameters of interest are land use, carbon sequestration, the introduction of pesticides and fertilizers, and the distance that the biomass needs to be transported to reach the plant site. The change of land can affect the environment directly and indirectly. When a forest or grassland is converted to crop land for biofuel generation, it is termed as "direct land-use change." Another type of changing of land usage when nonagricultural land is transformed into subsistence crop land as a result of land usage for biofuel production. This type of land-use change is termed "indirect land-use change" [36]. Most LCA analyses do not take a deeper look into the effects of land-use change; this de-emphasis on land-use change results in the under estimation of the carbon imbalance due to the displacement of subsistence crops. Kimming et al. recommended that some lands that are not used for food crops should not be used to harvest energy crops. This way, these set aside lands would hold carbon content in the soil [37]. The application of pesticides and fertilizers for energy crops means that their carbon footprint is much larger than their subsistence counterparts. Skowronska and Filipek's study found that a major portion of the 9.91 MJ energy used for the NPK (nitrogen, phosphorous, potassium) fertilizer is consumed by nitrogen. The technology and infrastructure are necessary to set up a fertilizer plant contribute to the increase in carbon dioxide, methane, and nitrogen oxide emissions [38]. Thus the use of agrochemicals is on the rise to address the food security challenges that we are facing due to limited land allocated to food crops. Consequently, the quality of the soil and the sequestered carbon worsen with the increasing use of agrochemicals. In this regards growing need to implement biomass waste from a fermentation plant or organic one to replace agrochemicals [38]. Agrochemicals have the adverse effect of removing essential organics matter from the soil as well. For instance, if enough straw residues are not left in the soil, an adverse impact on the soil's environment is expected. Removing straw inappropriately from agricultural land would impact the organic matter in the soil and reduce the biomass yield in the long run. This removal of straw residue from crop lands worsens ammonia's volatilization capability due to the greater mobility of fertilizers. Thus straw management is crucial because straw use is necessary for biofuel production and needs to be incorporated in LCA studies [39]. Transporting the biomass from the storage facility to the plant site is also a crucial step that needs to be incorporated. Most LCA studies in the literature have assumed that the plantation site is very close to the facility, thus offsetting the environmental burden caused due to transportation. The travel distance reported in the literature varied from 30 to 200 km. The transportation distance is strongly correlated with the size of the facility. The distance also strongly influences the overall GHG emissions of a certain thermochemical scheme. Phase 2 of this stage comprises pretreatment of biomass (grinding, crushing, and drying), thermochemical conversion, and upgradation of by-products. Pretreatment conditions for biomass vary a lot depending on the conditions. Moisture content and particle sizes after the pretreatment process solely depend on the requirements of the thermochemical process. Thus various conversion technologies contribute to the differences in environmental impact categories. Most studies in the literature have used pyrolysis, gasification, cofiring, and combustion as thermochemical conversion technologies. The material used for the equipment is also influenced by the process parameters, such as temperature, heating rate, and pressure. For instance, bio-oil is stored in a stainless-steel storage vessel. But crude oil-derived resources are necessary to obtain iron and manufacture stainless steel. This makes the stainless-steel vessel a big contributor to GHG emissions [40]. Thus the reduction of GHG emissions is a big challenge for the phase. The last phase is the demolition and recycling of the plant when its usage is no longer necessary. This phase involves the plant's demolition, metal extraction, transportation, and recycling. The materials which are not recyclable are landfilled. Due to the complexity involved, this phase is the least studied in previous LCA studies.

2.4.3 Inventory data collection

Data collection is the most demanding and time-consuming task in LCA studies. There are numerous methods to collect relevant data based on the budget and the resources available (Fig. 2.4). The data pertinent to LCA are classified into two major categories:

1. foreground data and
2. background data

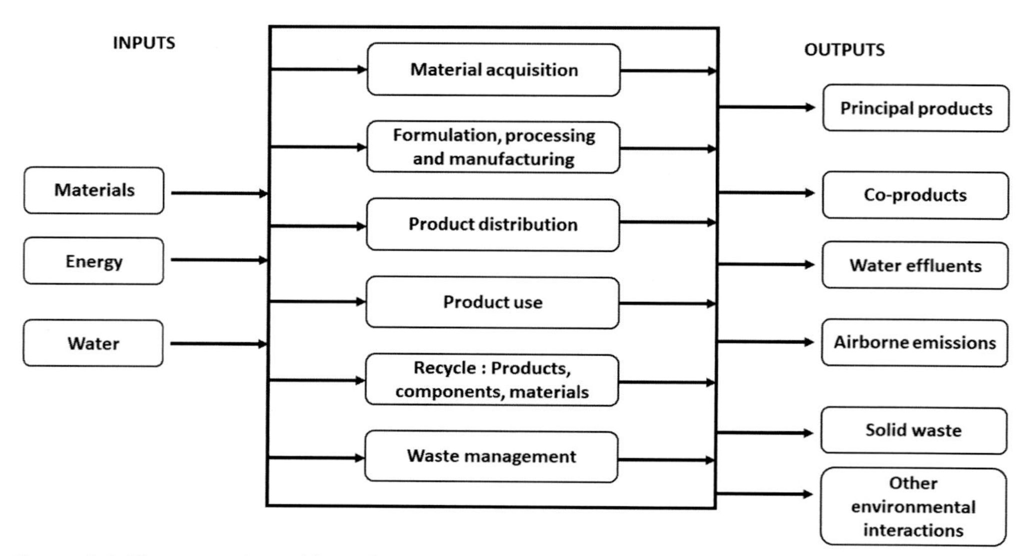

Figure 2.4 The stages in an life cycle assessment inventory analysis.

Foreground data is the set of specific data points relevant to the boundary. This data describes the products and their individual production systems. On the other hand, background data refers to the set of points describing transport, generic materials, waste management systems, and energy. Background data can usually be obtained by looking through databases or in the literature. However, data collected from other sources are not trivial. The following points should receive careful consideration for accurate results:

- The parties which possess background data will be encouraged to supply data if the researchers share a good relationship with them. Some companies will readily supply data; however other companies might see LCA studies as an impediment to their activities. In some instances, a major portion of the data collection effort establishes trust between the corporation and the researchers.
- The issues about breach of confidentiality might be very important.
- Each subset of an industry possesses its own unique terminology, a way to express and measure things. Thus if a researcher were to develop a questionnaire for a company, it might only apply to companies in a small subset of the industry.
- Questionnaires are generally the main modes of data collection. The questionnaire development should be handled with great responsibility as the questionnaire must connect well with the target company.

After the data collection and analysis stages are completed, mass balance and budget analyses are performed. A manufacturing process, shown in Fig. 2.5, has one of the simplest budget analysis methods. The process in the figure shows the cleaning of a product in a chemical process using a solvent. The scheme starts with the inclusion of a new solvent in the reservoir, which leads to a solvent displacement system to the product line where the solvent gets washed out. A major portion of the solvent is disposed into a recycling stream while a small amount is still retained in the product. This small portion is also known as "drag out." A little portion of this drag out stays on the product, while some of the material evaporates.

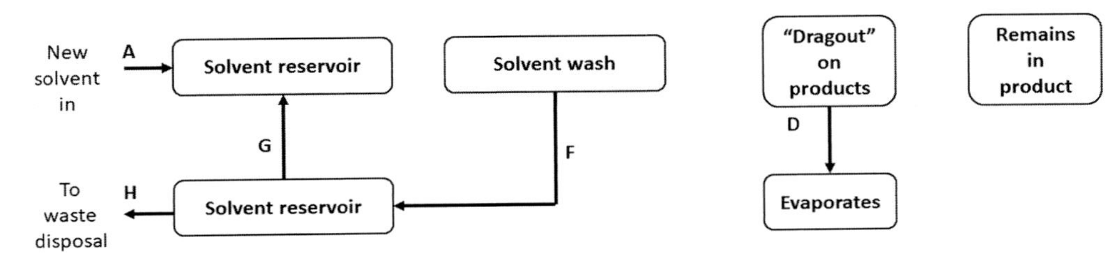

Figure 2.5 Schematic diagram of the flow streams involved in a budget analysis of a chemical solvent in a solvent washing process.

2.4.4 Analysis of life cycle inventory

The life cycle inventory analysis is the estimation of raw material and energy requirement, solid wastes, environmental emissions, water pollutants, and other emissions for the life of a process or product. Thus the result of an inventory analysis produces an estimation of the pollutants produced and the quantity of raw materials and energy consumed. To interpret the results, one can segregate the inventory analysis based on media, stage of life-cycle, life-cycle processes, or any arbitrary combination of these factors.

In general, the life cycle inventory phase involves the collection and organization of data. If a life cycle inventory did not exist, there is no basis to compare and evaluate different environmental impacts and suggest potential improvements.

The consequences of life cycle inventory are reflected downstream of the LCA process as well. Thus the accuracy of the collected data has substantial effects on the results produced by the LCA study.

In general, there are no predefined goals in terms of data quality for LCA analysis. Thus the data quality's nature depends on the accuracy necessary to provide insights into the decision process.

In some cases, the constraints on the resources can be considered during data collection; however, the scientific basis for LCA cannot tamper with it. The amount of information to be incorporated into an inventory is often influenced by the purpose and size of the study. For instance, in a study performed at a large scale, certain minute details can be left out as their contribution is insignificant for a study spanning across several industries. Thus these estimates may be left out without causing a significant change in the accuracy. However, in certain cases, these details may not be discounted. For instance, if a company wishes to compare various techniques or materials for inks used in packaging goods, then minute amounts of chemicals also need to be included. The inventory analysis of LCA study can be utilized in various ways. These data can aid an organization in efficiently evaluating the environmental impacts of certain decisions.

Furthermore, inventory analysis can be used in policy-making decisions. The data generated from the inventory analysis could aid the government in regulating resource use and application. Further insight into this matter can be gained by supplementing inventory analysis with an impact assessment study. For instance, we know that both carbon dioxide and methane emissions are harmful to the environment. However, to compare the individual impacts of these pollutants on global warming, smog, etc., an impact assessment can be useful.

2.4.5 Impact assessment of LCA

After setting the boundaries and collecting the inventory data, the next most important stage is the impact assessment of the LCA study. Eco indicator 99 and Eco indicator 95 are the preferred choices to estimate depletions in the ecosystem, human health, and other resources. Various environmental effect types can be estimated from the subdivisions of these categories. For instance, resources can be divided into energy utilization and solid wastes; human health can be subdivided into smog and other toxic pollutants, such as pesticides, carcinogens, and heavy metals; and ecosystem can be divided into acidification, eutrophication, and depletion of the ozone layer. The vast amount of literature in ecosystem impact assessment falls into the category of global warming potential. Other studies include assessments of eutrophication and acidification. The other categories (human health and resource depletion) have not been studied extensively in the literature. The CO_2 equivalent is used to measure the global warming potential; this equivalent estimates the CO_2, N_2O, and CH_4 emissions.

On the other hand, kg PO_2 and kg SO_2 are the standard equivalents used to estimate the eutrophication and acidification concentrations. Sebastián et al. concluded that plantation of biomass is much more benign in their production of GHGs than the application of pesticides. As a result of photosynthesis, the resultant carbon dioxide emissions are negligible during biomass cultivation. Sebastián et al. concluded that although energy crops are low maintenance, their impact on the environment is much worse than agricultural crops [41]. Pretreatment consumes a large portion of energy depending upon the mode of the thermochemical process. Biomass pretreatment comprises grinding, crushing, and drying. The energy consumption in the pretreatment process is strongly influenced by the moisture content of the feedstock while being inversely proportional to feedstock size. In general, agricultural residues have lower moisture content (<20% by wt.) than forest residues (40%–50% wt.) [28]. Therefore energy usage is dependent on the particle size of the feedstock. Iribarren et al. noted that during a fast pyrolysis process, the pretreatment of poplar wood had the most impact on the environment compared to other processes [42]. The distribution of particle sizes of lignocellulosic biomass is dependent on the process requirement. As the particle size increases, the char yield rises and the biofuel yield declines. Thus as the particles become much coarser, the desirable yield decreases [42].

Furthermore, a decrease in particle size for smaller particle sizes requires greater energy than larger particle sizes. For instance, reducing particle size from 25 to 3 mm requires 443 MJ/dry tonne of biomass, while a decrease in size from 300 to 25 mm required around 157 MJ/dry tonne. Environmental impact assessments are dependent on the operating parameters, such as pressure, the reactor's material, temperature, type of reactor, and heating rate. The GHG emissions are dependent on the construction of the plant and the material used. Implementing innovative technologies, such as biomass cofiring, would lead to a reduced impact on the environment. However, this technology also leads to a decline in the efficiencies of the boilers [43]. Rafaschieri et al. employed a pressurized fluidized bed gasifier to investigate the gasification process. Their study concluded that replacing oxygen as the oxidizer with air reduces the malignant effects on the environment for polar energy crops. This negative impact of using oxygen arises from the air separation process required to produce it; since this process is energy intensive and consumes electrical energy, the GHG emissions get increased [44].

2.4.6 Sensitivity analysis

The ranges of modeling information (various models and allocation methods), boundary, and activity data coupled with numerous end cases can make the accurate LCA of various plastics very difficult. Thus an uncertainty analysis is often performed to evaluate the progression of uncertainties in data and assumptions. This progression would determine the reliability of the data produced from the life cycle analysis [45]. The referenced ISO guidelines state that the uncertainty analysis results must complement the internal checks and balances setup within the evaluation procedure.

During assessments, ISO mandates assessments for consistency, sensitivity, and completeness [46]. To verify that LCA study is complete, gaps must be identified that could hinder the process of producing meaningful results. For instance, a waste water completeness check should include evaluating relevant life-cycle stages, such as operation and construction, and data quality. If the holes in these areas cannot be filled, the aim and the scope of the assessment must be reevaluated to address these gaps. Subsequently, sensitivity analysis should be conducted to quantify the dependence of the reliability of the results on the uncertainty [47,48]. The results generated through sensitivity analysis should be discussed during evaluation, and these sensitivity results should also motivate future data collection. After the results are examined, they must be checked again for consistency concerning the aim and scope of the LCA study [47,48]. For wastewater, an example of a consistency check is examining any spatial disconnect between the results and the sources.

Thus because of the amount and uncertainty of the data present in wastewater systems, sensitivity and uncertainty analysis should be conducted to examine the inferences usefulness. Within the framework of uncertainty analysis, Monte Carlo simulations, including Latin Hypercube sampling which drastically reduced the computation time, wherein input quantities are chosen randomly from the allocated datasets for about 10,000 simulations. Furthermore, the inputs in the LCA study, such as transportation distance and concentrations, are paired with probability distributions during uncertainty analyses to map the distributions in a simulated environment. A sampling technique known as Latin Hypercube [49] is increasingly being employed to produce results with good reproducibility and representation while reducing the number of simulations. Thus this technique reduces the required computational time by evenly sampling the input space. An important note for researchers is to include information about uncertainties while presenting data. A good way to include the uncertainty data is to report the ranges of quantified values.

On the other hand, sensitivity analysis is related to uncertainty analysis in the sense that sensitivity analysis encompasses the uncertainty relations between inputs and outputs. There may be various modalities, such as the one-at-a-time methods, variance-derived methods, or the correlation coefficient methods within sensitivity analysis. Popular examples of these modalities include the Morris method (one-at-a-time), Sobol method (variance-based), and Spearman's rank (correlation coefficient) methods. Wei et al. [50] provided extensive rules for selecting the right sensitivity modalities based on the type on the scope of the study and the model specifications. To add further detail to the sensitivity studies, scenario analyses are used to understand the scalability of conclusions and the influence of assumptions. For instance, as Dong et al. [51] demonstrated, the validity of conclusions can be checked, and different locations can be highlighted by assuming alternative electricity mixes. Studies performed by various research groups showed how the combined insights provided by uncertainty, sensitivity, and scenario analyses can be employed to make better decisions and identify data limitations [52—55].

2.5 Aspen plus approach to biomass pyrolysis system

The limited availability of time and the human errors present while manually handling complex problems on industrial applications make it highly challenging to solve them. Hence, the use of simulation programs is highly beneficial to predict the behavior of the process in such cases. One such simulation software package is Aspen Plus (Advanced System for Process Engineering), which is mainly used for the thermodynamic modeling of industrial processes. The package uses the fundamental engineering approach in accordance with the first law of thermodynamics (energy/mass balance) and kinetics of reactions to provide various features, including sensitivity analysis, unit operation models, and in-built templates. To simulate any process in Aspen Plus, the selection of property method is a crucial step for estimating the property of the compounds and pure substance. Numerous transport (viscosity, density, surface tension, etc.) and thermodynamic (enthalpy, thermal conductivity, Gibbs energy, etc.) properties are available as property methods in Aspen Plus. Aspen Plus also allows us to completely integrate the FORTRAN code or Excel subroutines as a flow sheet. It can simulate several complex industrial processes, such as coal gasification, biomass pyrolysis, fuel combustion, and gas and oil production. The basic process consists of mixing, reacting, heating, and cooling the given material in steady-state conditions and dividing unit operations [56—59]. The given compound at each stage of the process is moved from one unit to another through process streams. The volume or mass flow rates can be used to describe the stream of materials, while work or heat rates can be used to describe the energy stream. Material streams are broadly categorized into three types, which are called solid, mixed, and nonconventional streams. The mixed streams consist of a combination of many compounds and can be in any solid, liquid, or gas phase. The nonconventional material phase is used for materials like MSWs, biomass, coal, etc., and can be defined by their elemental composition and their heat of formation. Proximate and Ultimate analysis are some of the most commonly used methods to obtain the elemental composition of these nonconventional materials. Complex tasks, such as optimization, plant cost or physical properties estimation, and sensitivity analysis, can be performed through Aspen plus. Hence, the user must be aware of the behavior of the process and must be able to provide suitable process parameters to comprehend the results.

Gasification or pyrolysis process simulation through Aspen Plus can predict the characteristics and yield of the products obtained. Several studies have been conducted up to date on mathematical modeling of the pyrolysis and gasification process simulation. Once these studies have been implemented to produce a model, they must be calibrated and compared with the data obtained from the experimental investigations to validate their performance. When we require to simulate long processes [60] of thermochemical energy conversion, studies have shown that the thermodynamic equilibrium model (TEQ) is suitable for providing a good performance [61,62]. Hence, this model is beneficial for simulating the process of slow or intermediate pyrolysis and gasification (e.g., downdraft gasification [63—65]). One of the attributes of this model is that it does not consider the type, size, and shape of the reactor during the modeling process. Hence, it mainly uses the characteristics of biomass and the operating conditions to determine the efficiency of the process. It uses the energy and mass balance to calculate the product yield, composition of the yield products, and the heating value of the producer gas. The TEQ mathematical model can be created via stoichiometric or nonstoichiometric approaches. In stoichiometric modeling, it is necessary to know the chemical reactions undergone by the biomass during the thermochemical energy conversion process to calculate the value of equilibrium constants for the reactions [66,67]. The nonstoichiometric approach, on the other hand, uses the principle of Gibbs free energy and does not require the user to know the precise details of the chemical process [68].

Several studies have been conducted using Aspen Plus to simulate the thermochemical energy conversion of biomass [69,70], and each model is developed using a series of steps [71,72]. The first step is to identify the entire process phases. The next step is to represent each phase by selecting an appropriate model block. This is followed by establishing links between each model block, and the parameters for the process operation are defined. Since biomass falls under the category of the nonconventional material stream, it is characterized by entering the results from the ultimate and proximate analysis [73]. HCOALGEN and DCOALIG are tools on Aspen Plus, which can be used to determine the LHV, mass to volume ratio, and the heat/enthalpy of formation of the biomass. However, there is no single block available on Aspen Plus to represent the pyrolysis or gasification reactor. Hence, the user must model the reactor by utilizing two or more blocks. Multiple studies have modeled the gasification reactor using two blocks named R-yields and R-Gibbs [74,75]. The R-yields block does not consider the kinetics or stoichiometry of the reaction and simulates the process of volatilization occurring in the reactor. The R-Gibbs block determines the composition of the gas produced by minimizing the systems Gibbs free energy.

2.6 Kinetics of biomass pyrolysis

The kinetic modeling of a thermochemical conversion process requires us to determine the chemical reactions occurring in the process and find their kinetic constants $k(T)$. Commonly, thermogravimetric analysis (TGA) is used for calculating these $k(T)$, which in turn enables us to determine the values of activation energy (E_a) and the frequency or preexponential factor (A) in the Arrhenius equation [70,76–78]:

Using a single reaction, the biomass decomposition kinetics are found frequently, which can be delivered using the following canonical equation under isothermal conditions:

$$\frac{d\alpha}{dt} = k(T)f(\alpha) = A\exp\left(\frac{-E_a}{RT}\right)f(\alpha) \tag{2.1}$$

where $d\alpha/dt$ is the isothermal process rate, α denotes the conversion degree or amount of reaction, t represents time, and $f(\alpha)$ denotes the model of the reaction used as conversion function and dependent on the mechanism of control. The mass fraction of volatiles progressed, or the fraction of mass of decomposed biomass substrate can be stated as the degree of reaction, α, as shown below:

$$\alpha = \frac{w_0 - w}{w_0 - w_f} = \frac{v}{v_f} \tag{2.2}$$

where v_f is the reaction's product of the total mass of volatiles, v is the volatiles mass existing at any time t, w_f is the final solids mass (i.e., unreacted and residue substrate) balance at the end of the reaction, w_0 is the mass of initial substrate, and the mass of substrate present at any given moment t is denoted by w. The kinetic triplet is used to describe biomass pyrolysis reactions and combines the terms A, E_a, and $f(\alpha)$. Nonisothermal rate expressions describe the rate of reaction rates at a linear heating rate, β, as a function of temperature, which can be written through a superficial conversion of Eq. (2.1):

$$\frac{d\alpha}{dT} = \frac{d\alpha}{dt}\frac{dt}{dT} \tag{2.3}$$

where $d\alpha/dT$ denotes the rate of reaction of nonisothermal, $d\alpha/dt$ represents the reaction rate of isothermal, and dt/dT describes the inverse of the heating rate, $1/\beta$.

Replacing Eq. (2.1) with Eq. (2.3) yields the following equation of the rate law for nonisothermal conditions:

$$\frac{d\alpha}{dT} = \frac{k(T)}{\beta}f(\alpha) = \frac{A}{\beta}\exp\left(\frac{-E_a}{RT}\right)f(\alpha) \tag{2.4}$$

where R and T represent the ideal gas constant the absolute temperature, respectively.

Initially, the data for creating kinetic models of biomass pyrolysis had been obtained from the isothermal analysis. These required multiple experiments to be conducted at isothermal conditions, and due to this strenuous nature, the utilization of these methods has declined in recent days. Hence, dynamic nonisothermal techniques having the ability to perform investigations at different temperature ranges have garnered much attention lately [79]. These nonisothermal methods can monitor the biomass sample, which is subjected to a continuous increase in temperature, thereby collecting data at all points within the temperature range, thus beneficial over the isothermal techniques, which gather data at discrete temperatures. However, these nonisothermal techniques have been criticized and discredited by several authors [80–86]. The dynamic methods have shown a significant disparity in the obtained values in several studies, mainly due to the single heating rate being employed for the process. The lessons from the ICTAC commissioned kinetic projects [87–91] have shown a general agreement that the accuracy of these dynamic techniques can be enhanced by considering different groups of thermal data obtained at different heating rates.

The TGA method is a thermo analytical technique that determines the reduction in the mass due to volatilization of the biomass (solid-phase thermal degradation) concerning time or temperature. This method has been widely employed for assessing the thermal studies on biomass pyrolysis [92–95], and the data that must be input in the kinetic models are determined by taking the derivation of the TGA curve, which is called derivative thermogravimetry (DTG). From the DTG, the optimal temperature range for finding the kinetic parameters and the highest reaction rate [96] is found and applied as the input for the kinetic model. Besides TGA, differential scanning calorimetry (DSC) is another widely used technique to perform thermal analysis. The difference between DSC and DTA is that DTA is found at adiabatic conditions, allowing a temperature difference between the reference and the sample material. In DSC, reference materials, such as alumina and sample material, are simultaneously heated or cooled, and the heat flux that passes through them is compared. The heat of transition of the sample material used is directly proportional to the area obtained under the DSC curve.

Although thermo analytical analysis is an excellent method for assessing the reaction kinetics of heterogeneous reactions, no single method can accurately give conclusive results for the reaction mechanisms during the degradation process [97]. There is a possibility of change like the heterogeneous reaction system, which would represent the system accurately by just one specific kinetic model [98]. These changes occurring in the structure or composition of the sample material need to be detected and investigated by other analytical techniques. Evolved gas analysis is one such method that can be conducted using several analytical tools (Fourier-transform infrared spectroscopy (FTIR), Gas chromatography (GC), High performance liquid chromatography (HPLC), Gas chromatography-mass spectrometry (GC-MS), etc.) and provides an assessment of the gases that are evolved during the thermal degradation process. Hence, including the results from these additional methods in the thermal analysis can enable us to understand better the actual reaction mechanism occurring during the process.

The kinetic models widely used for biomass pyrolysis can be categorized into three different types: single-step global reaction, multistep, and semiglobal models [99—102]. The global single reaction techniques mainly describe the overall biomass volatilization rate. The single-step global reaction model has been validated by several researchers and has shown to demonstrate reasonably similar kinetic behavior when compared with results from an experimental investigation [103—107].

The biomass pyrolysis's devolatilization dynamics are commonly defined as a 1st order decomposition process that produces discrete volatile fractions:

$$\frac{dV_i}{dt} = k_i(T)\left(V_i^* - V_i\right) \tag{2.5}$$

Vi^* is the fraction i of the effective volatile content, Vi is the released volatiles cumulative mass corresponding to fraction i over time t, and $k_i(T)$ is the constant rate for an evolved volatile fraction i. The different volatilized fractions are graded in most devolatilization schemes for three biomass pseudo-components (i.e., lignin cellulose and hemicellulose) and, occasionally, moisture (Table 2.2).

However, these models work on the assumption that the mass ratio of the pyrolysis yield (oil, char, gas) is constant, which impedes its ability to forecast the pyrolysis yield based on reaction conditions. In addition, the complexity of the actual kinetics during the pyrolysis process makes it difficult to obtain the value of the apparent global activation energy [108].

Most of the work on modeling biomass pyrolysis has been conducted by the use of semiglobal models. They are used prominently as they enable us to perform a "lumped-system" kinetic analysis, which is relatively simpler than other types of analysis [92,100,109—112]. These models assume that the products from biomass pyrolysis can be distinctly categorized as chars, tars, and volatiles and intuitively determine the volatilization process as three concurrent first-order reactions. The parameters of the transport phenomenon and the secondary cracking reactions can be coupled using these models. These models have also been to provide accurate data for the trends in pyrolysis yield as a function of the residence time of volatiles [113]. It is a highly effective tool to evaluate the kinetics of different biomass materials when kept under similar

Table 2.2 Expressions for the most common reaction mechanisms in solid-state reactions.

Reaction model	$f(\alpha) = (1/k)(d\alpha/dt)$	$g(\alpha) = kt$
Order of reaction		
nth order	$(1-\alpha)^n$	$(n-1)^{-1}(1-\alpha)^{(1-n)}$
0th order	$(1-\alpha)^n$	α
1st order	$(1-\alpha)^n$	$-\ln(1-\alpha)$
Contracting geometry		
Contracting volume	$(1-\alpha)^{(1-1/n)}; n=3$	$1-(1-\alpha)^{1/n}; n=3$
Contracting area	$(1-\alpha)^{(1-1/n)}; n=2$	$1-(1-\alpha)^{1/n}; n=2$
Diffusional		
1D	$1/2\alpha$	α^2
2D	$[-\ln(1-\alpha)]^{-1}$	$(1-\alpha)\ln(1-\alpha)+\alpha$
3D (Ginstling–Brounshtein)	$3/2\left[(1-\alpha)^{-1/3}-1\right]^{-1}$	$1-2/3\alpha-(1-\alpha)^{2/3}$
3D (Jander)	$3/2(1-\alpha)^{2/3}\left[1-(1-\alpha)^{1/3}\right]^{-1}$	$\left[1-(1-\alpha)^{1/3}\right]^2$
Nucleation		
Prout–Tompkins	$\alpha(1-\alpha)$	$\ln\left[\alpha(1-\alpha)^{-1}\right]+C^a$
Avrami–Erofeev	$n(1-\alpha)[-\ln(1-\alpha)]^{(1-1/n)}; n=1,2,3,4$	$[-\ln(1-\alpha)]^{1/n}; n=1,2,3,4$
Exponential law	$\ln\alpha$	α
Power law	$n(\alpha)^{(1-1/n)}; n=2/3,1,2,3,4$	$\alpha^n; n=3/2,1,1/2,\frac{1}{3,\frac{1}{4}}$

process parameters but is unsuitable for evaluating the kinetic data obtained under dissimilar process parameters [105].

2.7 Isoconversional techniques

Solid-state processes, such as biomass pyrolysis, usually have convoluted reaction kinetics and undergo a complex set of concurrent and consecutive reactions. As mentioned earlier, initially, for each given set of isothermal data, iterative techniques were used for model-fitting and provide us with values for the Arrhenius parameters. Still, however, only a single set of the global kinetic triplet is obtained. Similarly, model-fitted isothermal data are used to consider the overall activation energy as global apparent activation energy is inadequate. Each stage in the process contains its own distinct value of apparent activation energy. This could also cause the DTG curves to mask the actual multistage pyrolysis process under a single peak [114].

Contrary to this, nonconsistent values for Arrhenius parameters and a strong dependency on the kinetic model are obtained on force-fitting models on a single heating rate nonisothermal data [115]. More reasonable values of kinetic parameters can be obtained by using different heating rates with nonisothermal techniques. However, since the extent to the dependency of heating rate varies for each reaction in the process, they can overlap one over the other in the DTG curves, which makes their separation demanding [116]. Hence, these difficulties paved the way for innovative research and development for creating models to determine suitable Arrhenius parameters for a reaction taking place with biomass species [115,117–121]. In the 1960s, a technique based on a single parameter named "model-free" was developed. It was an isoconversional method that assumed a constant degree of conversion for any reaction. This meant that the rate of the reaction, k, only depended upon the temperature of the reaction, T. These isoconversional methods remove the necessity of hypothesizing an initial rate order for the kinetic equation as they enable the calculation of E_a.

TGA data can be treated using either a differential or an integral method in isoconversional models. The Friedman method is a variance isoconversional method that can be written as follows in common terms:

$$\frac{d\alpha}{dt} = \beta\left(\frac{d\alpha}{dT}\right) = A\exp\left(\frac{-E_a}{RT}\right)f(\alpha) \tag{2.6}$$

Eq. (2.7) yields after taking natural logarithms on both sides from:

$$\ln\left(\frac{d\alpha}{dt}\right) = \ln\left[\beta\left(\frac{d\alpha}{dT}\right)\right] = \ln\left[Af(\alpha)\right] - \frac{E_a}{RT} \tag{2.7}$$

2.8 Other kinetic models

Numerous kinetic models besides the ones mentioned earlier have been developed, which supposedly offer improved results using biomass pyrolysis data. Balci et al. [122] had proposed a biomass deactivation model (BDM) based on the kinetic models, which are mainly used for catalyst deactivation purposes. In this model, the changes in the structure and chemical composition of the biomass during decomposition created a variation in the values of first-order rate constants. These rate constants were assumed to alter with the level of decomposition occurring during the pyrolysis process. This is because the composition of the reactive regions of the biomass substrate altered during pyrolysis as the individual components of biomass decomposed at different temperatures. The combined effect of reduction in volume, varied substrate geometry, and altering pore structure, which resulted in the loss of the total active surface area, was applied on both fossil [123–128] and plant-based [129–136] biomass pyrolysis. Another method called distributed activation energy model, I developed based on the surmise that numerous irreversible first-order reactions comprising of distinct kinetic parameters occur concurrently. The E_a from different reactions is shown by a continuous distribution function, $f(E_a)$, which, when approximated by a Gaussian distribution, provides the values for the standard deviation and mean of the activation energy. This method was revised in another study [137], which enabled the prediction of frequency factor and $f(E_a)$ without the necessity of any prior supposition of either of these parameters. The loss of weight curves from the coal pyrolysis process with multiple heating rates was predicted using this method. Another method to fit the nonisothermal data is the Weibull distribution model [138]. The model uses single or multiple Weibull distribution functions to represent the kinetic decomposition of each biomass substrate. The advantage of this model is that it allows us to deconvolute the overlapped processes from the TGA. Brown and Galwey mentioned that the presently used formal models are far too simple to evaluate a complex thermochemical phenomenon.

During pyrolysis, the deactivation of the solid has an impact on the apparent rate constant, as illustrated below:

$$k_{\text{app}} = zk = z(\gamma)\left(A_i\exp\left(-\frac{E_i}{RT}\right)\right) \tag{2.8}$$

where z denotes the behavior of the solid substrate as a function of a deactivation rate constant, γ, and k_{app} denotes the constant of apparent rate.

Reynolds et al. proposed a generalized model for nucleation growth, which is basically an alteration of the rate equation of Prout–Tompkins, which was first applied to potassium permanganate's thermal decomposition kinetics:

$$\frac{d\alpha}{dt} = ky^n(1-ry)^s \tag{2.9}$$

where r is the parameter of initiation frequently kept to 0.99, n is still the reaction order, y signifies the substrate's remaining fraction, and s represents a modifiable nucleation constraint, which can minimize Eq. (2.9) to 1st-order reaction, and the amount in parentheses $(1 - ry)$ substitutes $(1 - y)$ to keep the original rate from being 0. Traditional 1st-order models do not fit experimental data and this model, resulting in a substantially tighter deterioration curve.

2.9 Application of biomass pyrolysis products

Environment concern has highly developed in the past few decades due to the increase in fossil-based carbon. To reduce this, it is important to use a pyrolytic product to reduce the environmental issues and promote a country economically more substantially. This section will discuss the applications of the highly necessitated and evolved pyrolysis bio-oil and biochar.

2.9.1 Bio-oil applications

The different percentage of organic chemicals such as acids, alcohols, esters, aldehydes, phenol, alkanes, alkenes, ketones, sugar, and furans are various groups in bio-oil, which are derived from biomass are the main reason for the change various properties due to the variation of the chemical compounds in it as described in the previous section. After considering 300 different groups of chemicals present in the bio-oil leads to the determination of bio-oil applications.

2.9.1.1 Biochemicals

Levoglucosenone (LGO), levoglucosan (LG), furfural (FF), and phenolic compounds are some of the typical pyrolysis chemicals. LG, also known as (1,6-anhydrous-β-D-glucopyranose), is the most significant compound produced during the depolymerization reaction of pure cellulose or pyrolytic product of pure cellulose. The molecular formula for this compound is $C_6H_{10}O_5$, being used to produce bioethanol with hydrolyzed LG [139]. Using catalyst in the fast pyrolysis increases the chemical selectivity and product yield of LG [140], reacting 10% acetic acid increased 88% of the LG yield.

LGO, having the molecular formula $C_6H_6O_3$, is also known as (1,6-anhydrous-3,4-dideoxy-β-D-pyranosen-2-one). Similar to LG, LGO also a cellulose-derived product obtained from the process of depolymerization and dehydration reaction. Some other functional group can easily substitute the functional group present in this compound due to this characteristic. It is considered a promising compound, and this promising compound can be cast off for bioactive compounds, chiral inductors, tetrodotoxin, and disaccharides [141]. Using catalysts to increase the chemical selectivity and product yield of LGO, such as LG, on reacting 10% phosphoric acid (H_3PO_4) in pyrolysis enhanced the product yield of LGO to the maximum level of 62%. Cellulose fast pyrolysis with LGO forms 1-hydroxy-3,6-dioxabicyclo[1−3]octan2-one (LAC) is another form of C6 monomer. A large quantity of the LAC and LGO might be formed in bio-oil through an acid catalyst pyrolysis. The high boiling point characteristic of LG, LGO, and LGA prevents its production from bio-oil in simple distillation methods [142].

As discussed, another important chemical compound, FF, is produced from the compounds hemicellulose and cellulose. It has a wide range of applications in the area, such as resins, medicines, fuel, and plastics additives as an organic solvent. The liquid acid catalyst such as levoglucosenone (LGO) and levoglucosan (LG) helps to hydrolyze the xylose, fructose, and glucose 254 Waste Biorefinery in a further dehydration phase for FF production [143]. On using $MgCl_2$, it is observed that an FF yield of 80% was obtained. For instance, Wan et al. [144] for FF production used metal oxides, salt, and an acid catalyst of nine different types. Biomass lignin is used as the main source for pyrolytic lignin and monomeric phenolic compounds. Numerous reaches have examined a way to intensify the yield of phenols from biomass pyrolysis. Omoriyekomwan et al. [145] used palm kernel shells and produced almost 65% phenols in microwave pyrolysis bio-oil. We can produce exclusive and complex technologies to recuperate highly cleansed chemicals since they exist in small quantities. A corporation, Biomass Technology Group [146], produces monophenolic compound, pyrolytic lignin, organic acids, and pyrolytic sugars, which are all converted from bio-oil (derived from organic waste and biomass); those produced compounds are used as a substitute for the binder for asphalt product, including pyrolytic sugars, monophenolic production, and organic acids, a binder for asphalt, and a substitute for fossil phenol, and for fossil phenol pyrolytic lignin acts as a substitute.

2.9.1.2 Biofuel

Various examinations have occurred using fast pyrolysis biofuel in power generation and heat systems, including diesel engines, stirling engines, turbines, transportation fuels, and combustion burners. We can find many areas where fast pyrolysis bio-oil as a biofuel is used, such as a 5MWth swirl combustion furnace/burner that uses the bio-oil to burn; a business warmth and force plant (the United States, Wisconsin, Red Arrows) utilizing quick pyrolysis bio-oil was constructed in 1996 [147]. The internal combustion engine was examined to study the optimum operating conditions using pyrolysis oil [148]. Such investigations in the engine have pointed that using a large proportion of bio-oil causes a lag in the ignition period, which tends to the defective combustion of pyrolysis bio-oil [149]. Different research in the internal combustion engine has observed that the engine's operation is reliable by using a ratio of 6:4 upgraded bio-oil emulsified with biofuel [150]. To improve the ASTM standard of transport fuel, many researchers employ bio-oil as the fuel for transport. To increase the fuel properties, such as acid number, viscosity, water content, and heating value, it is must to increase the hydrogen content and reduce the oxygen content; to achieve these properties, the crude bio-oil is get upgraded through either catalytic hydrotreatment [151] or distillation [152]. Bio-oils are fractionated based on their relative volatility of chemicals at different temperatures using the physical upgrading method of distillation. On the other hand, hydrotreating and breaking C—C bonds through hydrocracking in the presence of hydrogen, the catalyst with the high pressure and high temperature reducing the oxygen content to improve the standard of bio-oil properties [153,151]. While upgrading the bio-oil process, some significant reactions, such as decarbonylation, hydrogenation, depolymerization, deoxygenation, and carbon accumulation, occur.

2.9.1.3 Biopolymer

For more than 50 years, biopolymers were produced from sources that are derived from fossils. The fast-growing environmental issues and fossil-fuel depletion have given thought to biodegradable polymers in society and scientists minds. Agro-resources, including polysaccharides, chitosan, and starch, are the main source of biopolymer production [154]. It is gathered that biopolymers not only act as environmentally pleasant but also formulate a way for material production using solid waste [155]. This part will see some examples for biopolymers, which include carbon fibers and phenolic resins processed from the pyrolysis oil.

There are numerous fields, such as computers, adhesives for wood, vehicles, and construction where this phenolic resin application is used [156]. The phenolic resin, which is part of the oldest thermoset polymers, is formed by the usual mixing of condensed petroleum-based phenol with formaldehyde. In the phenolic resin synthesis, the phenols present in it can be substituted by the pyrolytic bio-oil. In the manufacture of phenol-formaldehyde (PF) resin, higher viscosity and lower reactivity of bio-oil are caused due to the methoxy-substituted aromatic compound, which is the major reason phenol cannot be completely replaced by bio-oil. Sukhbaatar et al. [157] presented that a substitution rate of 30% yielded an equivalent PF performance linked with PF resin while using 40% and 50% bio-oil substitution gives growth to substandard performance. Chaouch et al. [158] by replacing phenol partially with the whole tree, bio-oil synthesized the phenolic resin. The performance of the parts replaced with phenolic resins exhibited equivalent or greater performance to the pure PF resins with virtuous thermal stability, shear strength, and storage stability.

Another biopolymer as carbon fiber for the last few decades is produced from the petroleum products, such as polyacrylonitrile (PAN). Since carbon fiber is an expensive material, many kinds of research were examined to condense the processing cost by possessing the fiber strength performance [159]. One of the prodigious intrant predecessors is lignin from lignocellulosic biomass solid wastes because of its low cost and renewable precursor. A study showed an average strength of 1.07 MPa, and Young's modulus of 82.7 GPa is good lignin-based carbon fiber [159]. There are several ways to produce carbon fiber in those approaches; melt spinning is the most common, which extrudes polymers through a small orifice in a metallic needle. After the lignin was desalted for extrusion, hardwood/softwood kraft lignin and oranganosolv hardwood lignin were used to fabricate the carbon fiber [160]. In the melt spinning process, hardwood lignin exhibited sophisticated thermal stability more than softwood lignin due to its crosslinking nature property. After treating organosoly hardwood to the carbonization and oxygen thermostabilization process, it showed Young's modulus of 40 GPa and strength of 400 MPa. Also, pyrolytic lignin derived from bio-oil is used to produce carbon fiber. Qin and Kadla [161] removed the water-soluble bio-oil available commercially to obtain pyrolytic lignin. For carbon fiber production pyrolytic lignin, the condensed bio-oil was undergone melt spun. Comparable to Alcell lignin and Kraft, the mechanical property of carbon fiber was found to be 36 GPa Young's modulus and 370 MPa strength. Qu et al. [162] used the melt-spinning method to produce pyrolytic lignin-based carbon fiber. Treating 0.5% sulfuric acid to pyrolytic lignin to further polymerize it gives a result that improves the property of carbon fiber includes greater molecular weight (DA). After treating the carbon fiber under the stabilization process, it is observed that it showed 855 MPa strength and 85 GPa Young's modulus, compared with acetylated softwood and Kraft hardwood lignin-based carbon fibers.

2.9.2 Biochar application

The biochar application is expanded in the various process as catalyst, energy storage device in the form of the electrode, and filter or adsorbent in the gas and liquid purification process—the biochar acts as both environmental and economically friendly.

2.9.2.1 Soil amendment

Biochar enhances the growth of plants by preventing the roots from getting affected by pathogens. Its large surface area causes the increase in soil aeration, which concedes the development of good soil bacteria to enhance the good supply of water and mineral. Since the biochar prevents certain soil problems, such as erosions, the soil contamination gets prevented due to this advantage, and the organisms are not affected by the toxicity's things [163]. The nature of the biochar, whether it is acidic or basic and hydrophobic or hydrophilic, is determined by the different surface functional groups present on it [164]. It is observed that hydrophobic natured biochar, when exposed to air and water for a long time, slowly turns into hydrophilic nature. This hydrophilic natured biochar is observed to have a great capacity for water retention, nutrient retention, and cation exchange property. These characteristics highly enhanced the production of crop and soil health.

On the other hand, many studies examined and reported the biochar does not hurt soil and crop growth [165]. The discussed effectiveness of biochar differs concerning soil properties, such as pH, feedstock, and surface area. Kambo and Dutta [166] when biochar positively influences the soil: When compared with the fertile soil, the working/effectiveness of biochar is observed to be good with unfertile soil, the effectiveness of biochar in temperate soil comparatively less when compared with tropical soil and in combination with nutrient-producing substances, and biochar having nutrient source is comparatively less operative compared with broiler litter biochar, which is very effective.

2.9.2.2 Solid biofuel

The energy density heating values of fossil fuel are most comparable with the biochar; therefore bio is simply used as a displacement for fossil fuel. The solid waste is processed to obtain biochar, and these biochars are found to have improved values, such as less energy required for pulverization, homogeneity, hydrophobicity, and high heating value. As an application for solid biofuel, torrefaction, also known as mild pyrolysis consider, that is, low-temperature pyrolysis around 200°C–300°C than the usual pyrolysis temperature around 400°C–600°C. For the rice straw, the range of HHV is from 19 to 28 MJ/kg [167], HHV for MSW is from 27 to 30 MJ/kg [168], and for

SS, it is from 26 to 31 MJ/kg [169]. The type of reactor and solid waste used will decide the HHV of the biochar, whether it is lower or higher compared to the torrefied biochar. Fig. 2.6 shows a Van Krevelen diagram based on the carbon, hydrogen, and oxygen composition of biochar elements, such as wood, de-oiled seed, algae, and rice straw. During the pyrolysis process, the decarboxylation and dehydration process/reaction with raw biomass loses hydrogen and oxygen, which leads to a change in the properties of biochar. The ratio of hydrogen and carbon and oxygen and carbon of torrefied biochar is found to be in the range of lignin.

In contrast, the ratio of those in biochar is in the range near bituminous coal. Therefore severe pyrolysis in biomass creates properties close to the range of coal. However, the torrefaction or mild pyrolysis allows solid waste to match the H/C and O/C ratio range close to lignin. The energy density is the ratio between the converted mass yield and the energy density of biochar, which increases if the biochar is

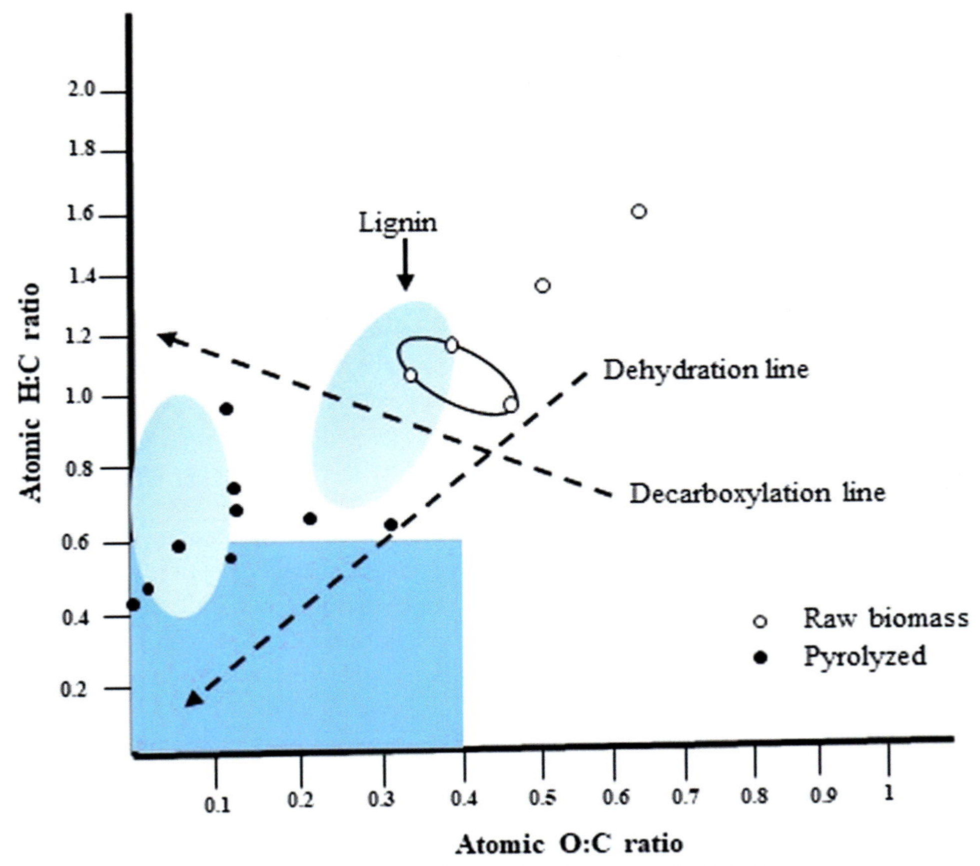

Figure 2.6 Van Krevelen graphical diagram for pyrolyzed and torrefied biochar.

processed from the improved biosolid waste. Nam et al. processed torrefied rice straw from the initial solid waste, which gives a weight reduction of 52% and 70% of energy recovery, which leads to obtain a maximum energy density of 1.48. The densified biochar acts economically friendly by reducing the transportation cost and storage difficulties.

2.9.2.3 Activated carbon

The production of porous carbon with surface area lies between 5 and 3000 m^2/g can be termed as the production of activated carbon [170], this activated carbon can be produced through chemical methods (e.g., HCL, KOH, and ZnCl2 activation) and physical methods (e.g., air activations, steam, and CO2). The chemical production method is dominant over the physical production method in the mass manufacture of activated carbon. The activated carbon produced from the chemical production method shows greater porosity and higher manageable porous property. The presents of pores in the activated carbon biochar will be classified according to its size i.e., pore size <20 A is micropore, $20 < \text{width} < 500$ A is mesopore and macropore one with size >500 A. Some of the applications of the characteristics of activated carbon include catalyst support, chemical purification adsorbent/filter, and electric power material.

The activated carbon as a purification filter/adsorbent removes some waste or polluted particles, such as dye, odor, and polluted particles present in air and water. As discussed, earlier volatile organic compounds (VOCs), Sulfur dioxide (SO_2), Hydrogen sulfide (H_2S), Nitrogen oxide (NOx), phenols, dyes, pesticides, heavy metals, and medicines are some of the contaminants. In the adsorption study of the pollutant removal process, pH plays a dominant role over additional parameters, such as initial contaminant concentration, process temperature, contact time, and adsorbent dose. The protonation propensity of the adsorbent will be high for less pH value or more acidic. Luna et al. [171] removed 96% of Eriochrome Black T dye at optimum condition using rice straw activated carbon with parameters of adsorbent dose, pH, and initial concentration. Mashayekh-Salehi and Moussavi [172] used pomegranate wood active carbon to remove a pollutant known to be acetaminophen from water. Following the Langmuir isotherm model and the pseudo-second-order kinetic model, the activated carbon achieved a maximum adsorption capacity of 233 mg/g.

Electrochemical materials, such as a battery, electric double-layer capacitor (EDLC), and supercapacitor, were from activated carbon derived from biomass. The EDLC and supercapacitor get differentiated by the reaction mechanisms. In an EDLC, only a pure electrostatic reaction will occur, whereas, in the supercapacitor, both electrostatic attraction and pseudocapacitance will occur. The biomass related to the production of electrochemical material was waste paper [173], wood sawdust [174], and cow dung [175]. Wei et al. [173] used wood sawdust active carbon and obtained a high specific capacitance of 236 F/g. Bhattacharjya and Yu [175] used natural waste cow dung activated carbon. They obtained a specific capacitance of 124 F/g, and Teo et al. [176] used KOH-activated carbon derived from the rice husk waste and used for a supercapacitor electrode. With 2695 m^2/g of surface area, 147 F/g was the highest specific capacitance.

The biochar-activated carbon can also be used as the catalyst. The main advantage of activated carbon as a catalyst is its high surface area, which can be changed by saturating metal catalysts. It is founded that the mesopore, which is <500 A present on the activated carbon, helps to observe the macromolecule chemical by penetrating through the pore, which is one of the major advantages for activated carbon as a catalyst [177]. Rajesh et al. [178] produced a Ni–Mo metal catalyst that is saturated on Jatropha-activated carbon and used in a hydrotreatment process. Compared with the Ni–Mo catalyst on alumina, the activated carbon as catalyst showed improved performance of bio-oil.

2.10 Conclusions

The pursuit for more environmentally sustainable and effective energy recovery waste management has prompted the improvement of pyrolysis-based waste to energy systems into the spotlight in recent years. However, current large-scale commercial plant case studies provide data that modern incineration can be an ecologically friendly technology that outperforms pyrolysis and gasification-melting plants. The pyrolysis-based waste to energy must develop the entire process flow to be economically effective (pretreatment, thermal conversion, product usage, residue management). The production of biofuel from generated biomass can contribute to the energy demand; it is critical to examine the environmental repercussions early on while the components of the system are still being established. The major sources of ecological impacts can be established. The consequent consequences can be avoided by examining the emissions, resource utilization, and energy use of the overall network, including biomass, transportation, and electricity generation. The LCA of a biomass power plant needs to be carried out for all these reasons, which included all of the plant's upstream production and downstream disposal operations. Various mathematical modeling models were developed to obtain the process and the model should be selected based on the parameters, operating conditions, requirements, and the economic benefit. Aspen plus is one of the significant simulation tools that help us to develop the economically feasible pyrolysis process in the near future as it is able to perform various tasks, such as optimization, plant cost or physical properties estimation, and sensitivity analysis. It helps to design and troubleshoot the process according to the behavior and the properties considered for the process.

References

[1] Dhyani V, Bhaskar T. A comprehensive review on the pyrolysis of lignocellulosic biomass. Renew Energy 2018;129:695–716.

[2] Ellabban O, Abu-Rub H, Blaabjerg F. Renewable energy resources: current status, future prospects and their enabling technology. Renew Sustain Energy Rev 2014;39:748–64.

[3] Vienescu DN, Wang J, Le Gresley A, Nixon JD. A life cycle assessment of options for producing synthetic fuel via pyrolysis. Bioresour Technol 2018;249:626–34.

[4] Alonso-Fariñas B, Gallego-Schmid A, Haro P, Azapagic A. Environmental assessment of thermo-chemical processes for bio-ethylene production in comparison with biochemical and fossil-based ethylene. J Clean Prod 2018;202:817–29.

[5] Ubando AT, Rivera DR, Chen WH, Culaba AB. A comprehensive review of life cycle assessment (LCA) of microalgal and lignocellulosic bioenergy products from thermochemical processes. Bioresour Technol 2019;291:121837.

[6] Zhao X, Zhou H, Sikarwar VS, Zhao M, Park AH, Fennell PS, et al. Biomass-based chemical looping technologies: the good, the bad and the future. Energy Environ Sci 2017;10(9):1885–910.

[7] Joseph K. Perspectives of solid waste management in India. In: Proceedings of the international symposium on the technology and management of the treatment & reuse of the municipal solid waste, Shanghai, China; 2002.

[8] Pamnani A, Srinivasarao M. Municipal solid waste management in India: a review and some new results. Int J Civ Eng Technol 2014;5(2):01–8.

[9] Imura H, Yedla S, Shirakawa H, Memon MA. Urban environmental issues and trends in Asia—an overview. Int Rev Environ Strateg 2005;5(2):357.

[10] Di Matteo U, Nastasi B, Albo A, Astiaso, Garcia D. Energy contribution of OFMSW (Organic Fraction of Municipal Solid Waste) to energy-environmental sustainability in urban areas at small scale. Energies 2017;10(2):229.

[11] Wang NY, Shih CH, Chiueh PT, Huang YF. Environmental effects of sewage sludge carbonization and other treatment alternatives. Energies 2013;6(2):871–83.

[12] Salata F, Golasi I, Domestico U, Banditelli M, Basso GL, Nastasi B, et al. Heading towards the nZEB through CHP + HP systems. A comparison between retrofit solutions able to increase the energy performance for the heating and domestic hot water production in residential buildings. Energy Convers Manage 2017;138:61–76.

[13] Rovense F, Amelio M, Ferraro V, Scornaienchi NM. Analysis of a concentrating solar power tower operating with a closed joule Brayton cycle and thermal storage. Int J Heat Technol 2016;34(3):485–90.

[14] Abubakar A, Barnabas MH, Tanko BM. The physico-chemical composition and energy recovery potentials of municipal solid waste generated in Numan Town. North Eastern Niger Energy Power Eng 2018;10(11):475–85.

[15] Ujam AJ, Eboh F. Thermal analysis of a small-scale municipal solid waste-fired steam generator: case study of Enugu State. Niger J Energy Technol Policy 2012;2(5):38–54.

[16] Arafat HA, Jijakli K, Ahsan A. Environmental performance and energy recovery potential of five processes for municipal solid waste treatment. J Clean Prod 2015;105:233–40.

[17] Kuleape R, Cobbina SJ, Dampare SB, Duwiejuah AB, Amoako EE, Asare W. Assessment of the energy recovery potentials of solid waste generated in Akosombo. Ghana Afr J Environ Sci Technol 2014;8(5):297–305.

[18] Management of municipal solid waste. New Delhi, India; 2018.

[19] Frassoldati A, Migliavacca G, Crippa T, Velata F, Faravelli T, Ranzi E. Detailed kinetic modeling of thermal degradation of biomasses. In: 29th meeting on combustion, Napoli, Italia: Combustion Institute; 2006.

[20] Pei-dong Z, Guomei J, Gang W. Contribution to emission reduction of CO_2 and SO_2 by household biogas construction in rural China. Renew Sustain Energy Rev 2007;11(8):1903–12.

[21] Thornley P, Upham P, Huang Y, Rezvani S, Brammer J, Rogers J. Integrated assessment of bioelectricity technology options. Energy Policy 2009;37(3):890–903.

[22] Chu S, Goldemberg J, Arungu Olende S, El-Ashry M, Davis G, Johansson T, et al. Lighting the way: toward a sustainable energy future. Amsterdam: Inter Academy Council; 2007.

[23] Grønli MG, Várhegyi G, Di Blasi C. Thermogravimetric analysis and devolatilization kinetics of wood. Ind Eng Chem Res 2002;41(17):4201–8.

[24] Fisher T, Hajaligol M, Waymack B, Kellogg D. Pyrolysis behavior and kinetics of biomass derived materials. J Anal Appl Pyrolysis 2002;62(2):331–49.

[25] Lanzetta M, Di Blasi C. Pyrolysis kinetics of wheat and corn straw. J Anal Appl Pyrolysis 1998;44(2):181–92.

[26] Jahirul MI, Rasul MG, Chowdhury AA, Ashwath N. Biofuels production through biomass pyrolysis—a technological review. Energies 2012;5(12):4952–5001.

[27] Venderbosch RH, Prins W. Biofuels. Bioprod Bioref 2010;4:178–208.

[28] Mann M, Spath P. A life cycle assessment of biomass cofiring in a coal-fired power plant. Clean Products Process 2001;3(2):81–91.

[29] Hsu DD. Life cycle assessment of gasoline and diesel produced via fast pyrolysis and hydroprocessing. Biomass Bioenergy 2012;45:41–7.

[30] Roberts KG, Gloy BA, Joseph S, Scott NR, Lehmann J. Life cycle assessment of biochar systems: estimating the energetic, economic, and climate change potential. Environ Sci Technol 2010;44:827–33.

[31] Koroneos C, Dompros A, Roumbas G. Hydrogen production via biomass gasification—a life cycle assessment approach. Chem Eng Process Process Intensif 2008;47(8):1261–8.

[32] Faix A, Schweinle J, Schöll S, Becker G, Meier D. (GTI-tcbiomass) life-cycle assessment of the BTO®-process (biomass-to-oil) with combined heat and power generation. Environ Prog Sustain Energy 2010;29(2):193–202.

[33] Lopez-Bellido L, Wery J, Lopez-Bellido RJ. Energy crops: prospects in the context of sustainable agriculture. Eur J Agron 2014;60:1–2.

[34] Kumar A, Cameron JB, Flynn PC. Biomass power cost and optimum plant size in western Canada. Biomass Bioenergy 2003;24(6):445–64.

[35] Nguyen TL, Hermansen JE, Mogensen L. Environmental performance of crop residues as an energy source for electricity production: the case of wheat straw in Denmark. Appl Energy 2013;104:633–41.

[36] Muralikrishna IV, Manickam V. Environmental management life cycle assessment. Environmental management. Berlin, Germany: Springer; 2017. p. 57–75.

[37] Lange M. The GHG balance of biofuels taking into account land use change. Energy Policy 2011;39(5):2373–85.

[38] Kimming M, Sundberg C, Nordberg Å, Baky A, Bernesson S, Norén O, et al. Biomass from agriculture in small-scale combined heat and power plants—a comparative life cycle assessment. Biomass Bioenergy 2011;35(4):1572–81.

[39] Skowrońska M, Filipek T. Life cycle assessment of fertilizers: a review. Int Agrophys 2014;28(1):101–10.

[40] Gabrielle B, Gagnaire N. Life-cycle assessment of straw use in bio-ethanol production: A case study based on biophysical modelling. Biomass Bioenergy 2008;32(5):431–41.

[41] Martínez E, Sanz F, Pellegrini S, Jiménez E, Blanco J. Life cycle assessment of a multi-megawatt wind turbine. Renew Energy 2009;34(3):667–73.

[42] Sebastián F, Royo J, Gómez M. Cofiring vs biomass-fired power plants: GHG (greenhouse gases) emissions savings comparison by means of LCA (life cycle assessment) methodology. Energy 2011;36(4):2029–37.

[43] Iribarren D, Peters JF, Dufour J. Life cycle assessment of transportation fuels from biomass pyrolysis. Fuel 2012;97:812–21.

[44] Rafaschieri A, Rapaccini M, Manfrida G. Life cycle assessment of electricity production from poplar energy crops compared with conventional fossil fuels. Energy Convers Manage 1999;40(14):1477–93.

[45] ISO. Environmental management—life cycle assessment—requirements and guidelines. ISO 14044. Geneva; 2006.

[46] ISO. Environmental management—life cycle assessment. Principles and framework. ISO 14040. Geneva; 2006, p. 2006.

[47] ISO 14044. Environmental management—life cycle assessment—requirements and guidelines. Geneva; 2006.
[48] ISO 14040. Environmental management—life cycle assessment—principles and framework: international. Standard 14040. Geneva; 2006.
[49] McKay MD, Beckman RJ, Conover WJ. A comparison of three methods for selecting values of input variables in the analysis of output from a computer code. Technometrics 2000;42(1):55−61.
[50] Wei W, Larrey-Lassalle P, Faure T, Dumoulin N, Roux P, Mathias JD. How to conduct a proper sensitivity analysis in life cycle assessment: taking into account correlations within LCI data and interactions within the LCA calculation model. Environ Sci Technol 2015;49(1):377−85.
[51] Dong S, Li J, Kim M-H, Park S-J, Eden JG, Guest JS, et al. Human health trade-offs in the disinfection of wastewater for landscape irrigation: microplasma ozonation vs. chlorination. Environ Sci Water Res Technol 2017;3(1):106−18.
[52] Foley JM, Rozendal RA, Hertle CK, Lant PA, Rabaey K. Life cycle assessment of high-rate anaerobic treatment, microbial fuel cells, and microbial electrolysis cells. Environ Sci Technol 2010;44 (9):3629−37.
[53] Foley J, de Haas D, Hartley K, Lant P. Comprehensive life cycle inventories of alternative wastewater treatment systems. Water Res 2010;44(5):1654−66.
[54] Niero M, Pizzol M, Bruun HG, Thomsen M. Comparative life cycle assessment of wastewater treatment in Denmark including sensitivity and uncertainty analysis. J Clean Prod 2014;68:25−35.
[55] Shoener BD, Zhong C, Greiner AD, Khunjar WO, Hong P, Guest JS. Design of anaerobic membrane bioreactors for the valorization of dilute organic carbon waste streams. Energy Environ Sci 2016;9(3):1102−12.
[56] Prabowo B, Aziz M, Umeki K, Susanto H, Yan M, Yoshikawa K. CO_2-recycling biomass gasification system for highly efficient and carbon-negative power generation. Appl Energy 2015;158:97−106.
[57] Rahman A, Rasul MG, Khan MMK, Sharma S. Aspen plus based simulation for energy recovery from waste to utilize in cement plant preheater tower. Energy Procedia 2014;61:922−7.
[58] Ong'iro A, Ugursal VI, Al Taweel AM, Lajeunesse G. Thermodynamic simulation and evaluation of a steam CHP plant using ASPEN Plus. Appl Therm Eng 1996;16(3):263−71.
[59] Pala LPR, Wang Q, Kolb G, Hessel V. Steam gasification of biomass with subsequent syngas adjustment using shift reaction for syngas production: an Aspen Plus model. Renew Energy 2017;101:484−92.
[60] Baratieri M, Baggio P, Fiori L, Grigiante M. Biomass as an energy source: thermodynamic constraints on the performance of the conversion process. Bioresour Technol 2008;99(15):7063−73.
[61] Syed S, Janajreh I, Ghenai C. Thermodynamics equilibrium analysis within the entrained flow gasifier environment. Int J Therm Environ Eng 2011;4(1):47−54.
[62] Melgar A, Pérez JF, Laget H, Horillo A. Thermochemical equilibrium modelling of a gasifying process. Energy Convers Manage 2007;48(1):59−67.
[63] Baruah D, Baruah DC. Modeling of biomass gasification: a review. Renew Sustain Energy Rev 2014;39:806−15.
[64] Mhilu CF. Modeling performance of high-temperature biomass gasification process. ISRN Chem Eng 2012;2012:1−13.
[65] Gagliano A, Nocera F, Patania F, Bruno M. A numerical model to characterize the producer gas composition by the pyrolysis process. In: IREC 7th International Renewable Energy Congress, vol. Dii; 2016. p. 31−36.
[66] Zainal ZA, Ali R, Lean CH, Seetharamu KN. Prediction of performance of a downdraft gasifier using equilibrium modeling for different biomass materials. Energy Convers Manage 2001;42 (12):1499−515.
[67] Jarungthammachote S, Dutta A. Thermodynamic equilibrium model and second law analysis of a downdraft waste gasifier. Energy 2007;32(9):1660−9.
[68] Altafini CR, Wander PR, Barreto RM. Prediction of the working parameters of a wood waste gasifier through an equilibrium model. Energy Convers Manage 2003;44(17):2763−77.
[69] Doherty W, Reynolds A, Kennedy D. The effect of air preheating in a biomass CFB gasifier using ASPEN Plus simulation. Biomass Bioenergy 2009;33(9):1158−67.

[70] Doherty W, Reynolds A, Kennedy D. Simulation of a circulating fluidised bed biomass gasifier using ASPEN Plus—A performance analysis. In: Ecos. Proceedings of the 21st international conference on efficiency, cost, optimization, simulation and environmental impact of energy systems; 2008, p. 1241−48.
[71] Begum S, Rasul MG, Akbar D. A numerical investigation of municipal solid waste gasification using aspen plus. Proc Eng 2014;90:710−17.
[72] Miccio VM, Juchelková D. An aspen plus® tool for simulation of lignocellulosic biomass pyrolysis via equilibrium and ranking of the main process variables. Int J Math Model Methods Appl Sci 2015;9:71−86.
[73] Adeyemi JI. Modeling of the entrained flow gasification: kinetics-based ASPEN Plus model. 2014:1−8.
[74] Kabir MJ, Chowdhury AA, Rasul MG. Pyrolysis of municipal green waste: a modelling, simulation and experimental analysis. Energies 2015;8(8):7522−41.
[75] Ward J, Rasul MG, Bhuiya MMK. Energy recovery from biomass by fast pyrolysis. Proc Eng 2014;90:669−74.
[76] White JE, Catallo WJ, Legendre BL. Biomass pyrolysis kinetics: a comparative critical review with relevant agricultural residue case studies. J Anal Appl Pyrol 2011;91(1):1−33.
[77] Roy PC, Datta A, Chakraborty N. An assessment of different biomass feedstocks in a downdraft gasifier for engine application. Fuel 2013;106:864−8.
[78] Jeya Singh VC, Sekhar SJ. Performance studies on a downdraft biomass gasifier with blends of coconut shell and rubber seed shell as feedstock. Appl Therm Eng 2016;97:22−7.
[79] Agrawal RK. Analysis of non-isothermal reaction kinetics. Part 1. Thermochim Acta 1992;203 (C):93−110.
[80] Tang TB, Chaudhri MM. Analysis of dynamic kinetic data from solid-state reactions. J Therm Anal 1980;18(2):247−61.
[81] Rasool SI, de Bergh C. © 1970 Nature Publishing Group, vol. 228; 1970. p. 726−34. Available from: http://www.mendeley.com/research/discreteness-conductance-chnge-n-bimolecular-lipid-membrane-presence-certin-antibiotics/.
[82] Garn PD. Introduction and critique of non-isothermal kinetics. Thermochim Acta 1987;110(C):141−4.
[83] Howell BA, Ray JA. Comparison of isothermal and dynamic methods for the determination of activation energy by thermogravimetry. J Therm Anal Calorim 2006;83(1):63−6.
[84] Norwisz JAN. Of the reaction, or during a non-isothermal reaction run. There is only one value x = b for each point P(T, 2,). The differential equation describing the kinetics. p. 4−6.
[85] Cooney JD, Day M, Wiles DM. Thermal degradation of poly(ethylene terephthalate): a kinetic analysis of thermogravimetric data. J Appl Polym Sci 1983;28(9):2887−902. Available from: https://doi.org/10.1002/app.1983.070280918.
[86] TJi-IERMOGRAVIMETRIC DATA. Science. 1994;7(94):61−67.
[87] Vyazovkin S. Computational aspects of kinetic analysis. Thermochim Acta 2000;355(1−2):155−63.
[88] Brown ME, Maciejewski M, Vyazovkin S, Nomen R, Sempere J, Burnham A, et al. Computational aspects of kinetic analysis. Thermochim Acta 2000;355(1−2):125−43.
[89] Maciejewski M. Computational aspects of kinetic analysis. Thermochim Acta 2000;355(1−2):145−54.
[90] Roduit B. Computational aspects of kinetic analysis. Thermochim Acta 2000;355(1−2):171−80.
[91] Burnham AK. Computational aspects of kinetic analysis. Thermochim Acta 2000;355(1−2):165−70.
[92] Branca C, Iannace A, di Blasi C. Devolatilization and combustion kinetics of Quercus cerris bark. Energy Fuels 2007;21(2):1078−84.
[93] Mészáros G, Jakab VE, Marosvölgyi B. Thermogravimetric and reaction kinetic analysis of biomass samples from an energy plantation. Energy Fuels 2004;18(2):497−507.
[94] Stenseng M, Jensen A, Dam-Johansen K. Investigation of biomass pyrolysis by thermogravimetric analysis and differential scanning calorimetry. J Anal Appl Pyrol 2001;58−59:765−80.
[95] Teng H, Lin HC, Ho JA. Thermogravimetric analysis on global mass loss kinetics of rice hull pyrolysis. Ind Eng Chem Res 1997;36(9):3974−7.
[96] Kissinger HE. Reaction kinetics in differential thermal analysis. Anal Chem 1957;29(11):1702−6.
[97] Howell BA. Utility of kinetic analysis in the determination of reaction mechanism. J Therm Anal Calorim 2006;85(1):165−7. Available from: https://doi.org/10.1007/s10973-005-7484-z.

[98] Brown ME, Galwey AK. The distinguishability of selected kinetic models for isothermal solid-state reactions. Thermochim Acta 1979;29(1):129–46. Available from: https://doi.org/10.1016/0040-6031(79)85024-8.
[99] Hajaligol MR, Howard JB, Longwell JP, Peters WA. Product compositions and kinetics for rapid pyrolysis of cellulose. Ind Eng Chem Proc Des Dev 1982;21(3):457–65. Available from: https://doi.org/10.1021/i200018a019.
[100] Thurner F, Mann U. Kinetic investigation of wood pyrolysis. Ind Eng Chem Proc Des Dev 1981;20(3):482–8. Available from: https://doi.org/10.1021/i200014a015.
[101] Tsamba AJ, Yang W, Blasiak W. Pyrolysis characteristics and global kinetics of coconut and cashew nut shells. Fuel Process Technol 2006;87(6):523–30.
[102] Agrawal RK. Kinetics of reactions involved in pyrolysis of cellulose I. The three reaction model. Can J Chem Eng 1988;66(3):403–12.
[103] Nunn TR, Howard JB, Longwell JP, Peters WA. Product compositions and kinetics in the rapid pyrolysis of milled wood lignin. Ind Eng Chem Proc Des Dev 1985;24(3):844–52.
[104] Villermaux J, Antoine B, Lede J, Soulignac F. A new model for thermal volatilization of solid particles undergoing fast pyrolysis. Chem Eng Sci 1986;41(1):151–7.
[105] Nunn TR, Howard JB, Longwell JP, Peters WA. Product compositions and kinetics in the rapid pyrolysis of sweet gum hardwood. Ind Eng Chem Proc Des Dev 1985;24(3):836–44.
[106] Cordero T, García F, Rodríguez JJ. A kinetic study of holm oak wood pyrolysis from dynamic and isothermal TG experiments. Thermochim Acta 1989;149:225–37.
[107] Antal MJ, Várhegyi G, Jakab E. Cellulose pyrolysis kinetics: revisited. Ind Eng Chem Res 1998;37(4):1267–75. Available from: https://doi.org/10.1021/ie970144v.
[108] Flynn JH. Thermal analysis kinetics-problems, pitfalls and how to deal with them. J Therm Anal 1988;34(1):367–81.
[109] Di Blasi C. Kinetic and heat transfer control in the slow and flash pyrolysis of solids. Ind Eng Chem Res 1996;35(1):37–46.
[110] Shafizadeh F, Chin PPS. Thermal deterioration of wood; 1977.
[111] Radmanesh R, Courbariaux Y, Chaouki J, Guy C. A unified lumped approach in kinetic modeling of biomass pyrolysis. Fuel 2006;85(9):1211–20.
[112] di Blasi C. Comparison of semi-global mechanisms for primary pyrolysis of lignocellulosic fuels. J Anal Appl Pyrol 1998;47(1):43–64.
[113] Chan WR, Kelbon M, Krieger BB. Modelling and experimental verification of physical and chemical processes during pyrolysis of a large biomass particle. Fuel 1985;64(11):1505–13.
[114] Mamleev V, Bourbigot S. Calculation of activation energies using the sinusoidally modulated temperature. J Therm Anal Calorim 2002;70(2):565–79.
[115] Vyazovkin S, Wight CA. Isothermal and non-isothermal kinetics of thermally stimulated reactions of solids. Int Rev Phys Chem 1998;17(3):407–33.
[116] Cao R, Naya S, Artiaga R, García A, Varela A. Logistic approach to polymer degradation in dynamic TGA. Polym Degrad Stab 2004;85(1):667–74.
[117] Budrugeac P. Differential non-linear isoconversional procedure for evaluating the activation energy of non-isothermal reactions. J Therm Anal Calorim 2002;68(1):131–9.
[118] Li CR, Tang TB. Isoconversion method for kinetic analysis of solid-state reactions from dynamic thermoanalytical data. J Mater Sci 1999;34(14):3467–70.
[119] Wanjun T, Donghua C. An integral method to determine variation in activation energy with extent of conversion. Thermochim Acta 2005;433(1–2):72–6.
[120] Ortega A. A simple and precise linear integral method for isoconversional data. Thermochim Acta 2008;474(1–2):81–6.
[121] Han Y, Chen H, Liu N. New incremental isoconversional method for kinetic analysis of solid thermal decomposition. J Therm Anal Calorim 2011;104(2):679–83.
[122] Balcı S, Doğu T, Yücel H. Pyrolysis kinetics of lignocellulosic materials. Ind Eng Chem Res 1993;32(11):2573–9.
[123] Agarwal P. Distributed kinetic parameters for methane evolution during coal pyrolysis. Fuel 1985;64(6):870–2.

[124] Anthony DB, Howard JB, Hottel HC, Meissner HP. Rapid devolatilization of pulverized coal. Symp (Int) Combust 1975;15(1):1303–17.
[125] Anthony DB, Howard JB. Coal devolatilization and hydrogasification. AIChE J 1976.
[126] Biagini E, Lippi F, Petarca L, Tognotti L. Devolatilization rate of biomasses and coal-biomass blends: an experimental investigation. Fuel 2002;81(8):1041–50.
[127] Braun RL, Burnham AK. Analysis of chemical reaction kinetics using a distribution of activation energies and simpler models. Energy Fuels 1987;1(2):153–61.
[128] Agarwal PK, Agnew JB, Ravindran N, Weimann R. Distributed kinetic parameters for the evolution of gaseous species in the pyrolysis of coal. Fuel 1987;66(8):1097–106.
[129] Várhegyi G, Szabó P, Antal MJ. Kinetics of charcoal devolatilization. Energy Fuels 2002;16 (3):724–31.
[130] Várhegyi G, Czégény Z, Jakab E, McAdam K, Liu C. Tobacco pyrolysis. Kinetic evaluation of thermogravimetric-mass spectrometric experiments. J Anal Appl Pyrolysis 2009;86(2):310–22.
[131] Wang G, Li W, Li B, Chen H. TG study on pyrolysis of biomass and its three components under syngas. Fuel 2008;87(4–5):552–8.
[132] Sonobe T, Worasuwannarak N. Kinetic analyses of biomass pyrolysis using the distributed activation energy model. Fuel 2008;87(3):414–21.
[133] Becidan M, Várhegyi G, Hustad JE, Skreiberg Ø. Thermal decomposition of biomass wastes. A kinetic study. Ind Eng Chem Res 2007;46(8):2428–37.
[134] Várhegyi G, Chen H, Godoy S. Thermal decomposition of wheat, oat, barley, and brassica carinata straws. a kinetic study. Energy Fuels 2009;23(2):646–52.
[135] Boroson ML, Howard JB, Longwell JP, Peters WA. Product yields and kinetics from the vapor phase cracking of wood pyrolysis tars. AIChE J 1989;35(1):120–8.
[136] Tsamba AJ, Yang W, Blasiak W, Wójtowicz MA. Cashew nut shells pyrolysis: individual gas evolution rates and yields. Energy Fuels 2007;21(4):2357–62.
[137] Miura K, Maki T. A simple method for estimating f(E) and k0(E) in the distributed activation energy model. Energy Fuels 1998;12(5):864–9.
[138] Cai J, Liu R. Weibull mixture model for modeling nonisothermal kinetics of thermally stimulated solid-state reactions: application to simulated and real kinetic conversion data. J Phys Chem B 2007;111(36):10681–6.
[139] Meng X, Zhang H, Liu C, Xiao R. Comparison of acids and sulfates for producing levoglucosan and levoglucosenone by selective catalytic fast pyrolysis of cellulose using Py-GC/MS. Energy Fuels 2016;30(10):8369–76.
[140] Bennett NM, Helle SS, Duff SJ. Extraction and hydrolysis of levoglucosan from pyrolysis oil. Bioresour Technol 2009;100(23):6059–63.
[141] Sarotti AM. Theoretical insight into the pyrolytic deformylation of levoglucosenone and isolevoglucosenone. Carbohydr Res 2014;390:76–80.
[142] Zhu XF, Lu Q. Production of chemicals from selective fast pyrolysis of biomass. Croatia: Sciyo; 2010.
[143] Chheda JN, Román-Leshkov Y, Dumesic JA. Production of 5-hydroxymethylfurfural and furfural by dehydration of biomass-derived mono-and poly-saccharides. Green Chem 2007;9(4):342–50.
[144] Wan Y, Chen P, Zhang B, Yang C, Liu Y, Lin X, et al. Microwave-assisted pyrolysis of biomass: catalysts to improve product selectivity. J Anal Appl Pyrolysis 2009;86(1):161–7.
[145] Omoriyekomwan JE, Tahmasebi A, Yu J. Production of phenol-rich bio-oil during catalytic fixedbed and microwave pyrolysis of palm kernel shell. Bioresour Technol 2016;207:188–96.
[146] Biomass Technology Group. Technologies bio-materials & chemicals.
[147] Adhikari S, Nam H, Chakraborty JP. Conversion of solid wastes to fuels and chemicals through pyrolysis. Waste Biorefinery 2018;239–63.
[148] Chiaramonti D, Oasmaa A, Solantausta Y. Power generation using fast pyrolysis liquids from biomass. Renew Sustain Energy Rev 2007;11(6):1056–86.
[149] Shihadeh A, Hochgreb S. Diesel engine combustion of biomass pyrolysis oils. Energy Fuels 2000;14(2):260–74.
[150] Laesecke J, Ellis N, Kirchen P. Production, analysis and combustion characterization of biomass fast pyrolysis oil—Biodiesel blends for use in diesel engines. Fuel 2017;199:346–57.

[151] Nam H, Kim C, Capareda SC, Adhikari S. Catalytic upgrading of fractionated microalgae bio-oil (Nannochloropsis oculata) using a noble metal (Pd/C) catalyst. Algal Research 2017;24:188–98.
[152] Nam H, Choi J, Capareda SC. Comparative study of vacuum and fractional distillation using pyrolytic microalgae (Nannochloropsis oculata) bio-oil. Algal Research 2016;17:87–96.
[153] Wang Z, Adhikari S, Valdez P, Shakya R, Laird C. Upgrading of hydrothermal liquefaction biocrude from algae grown in municipal wastewater. Fuel Process Technol 2016;142:147–56.
[154] Averous L, Pollet E. Environmental silicate nano-biocomposites. London: Springer; 2012.
[155] Pacheco-Torgal F. Introduction to biopolymers and biotech admixtures for eco-efficient construction materials. Biopolymers and biotech admixtures for eco-efficient construction materials. Woodhead Publishing; 2016. p. 1–10.
[156] Cardona F, Kin-Tak AL, Fedrigo J. Novel phenolic resins with improved mechanical and toughness properties. J Appl Polym Sci 2012;123(4):2131–9.
[157] Sukhbaatar B, Steele PH, Ingram LI, Kim MG. Use of lignin separated from bio-oil in oriented strand board binder phenol-formaldehyde resins. BioResources 2009;4(2):789–804.
[158] Chaouch M, Diouf PN, Laghdir A, Yin S. Bio-oil from whole-tree feedstock in resol-type phenolic resins. J Appl Polym Sci 2014;131(6).
[159] Baker DA, Rials TG. Recent Advances in Low-Cost Carbon Fiber Manufacture from Lignin. J Appl Polym Sci 2013;130(2):713–28.
[160] Kadla JF, Kubo S, Venditti RA, Gilbert RD, Compere AL, Griffith W. Lignin-based carbon fibers for composite fiber applications. Carbon 2002;40(15):2913–20.
[161] Qin W, Kadla JF. Carbon fibers based on pyrolytic lignin. J Appl Polym Sci 2012;126(S1): E204–13.
[162] Qu W, Xue Y, Gao Y, Rover M, Bai X. Repolymerization of pyrolytic lignin for producing carbon fiber with improved properties. Biomass Bioenergy 2016;95:19–26.
[163] Beesley L, Marmiroli M. The immobilisation and retention of soluble arsenic, cadmium and zinc by biochar. Environ Pollut 2011;159(2):474–80.
[164] Amonette JE, Joseph S. Characteristics of biochar: microchemical properties. Biochar for environmental management. Routledge; 2012. p. 65–84.
[165] Jeffery S, Verheijen FGA, Van Der Velde M, Bastos AC. A quantitative review of the effects of biochar application to soils on crop productivity using meta-analysis. Agric Ecosyst Environ 2011;144(1):175–87.
[166] Kambo HS, Dutta A. A comparative review of biochar and hydrochar in terms of production, physico-chemical properties and applications. Renew Sustain Energy Rev 2015;45:359–78.
[167] Nam H, Capareda S. Experimental investigation of torrefaction of two agricultural wastes of different composition using RSM (response surface methodology). Energy 2015;91:507–16.
[168] Yuan H, Wang Y, Kobayashi N, Zhao D, Xing S. Study of fuel properties of torrefied municipal solid waste. Energy Fuels 2015;29(8):4976–80.
[169] Dhungana A, Dutta A, Basu P. Torrefaction of non-lignocellulose biomass waste. Can J Chem Eng 2012;90(1):186–95.
[170] Rouquerol J, Rouquerol F, Llewellyn P, Maurin G, Sing KS. Adsorption by powders and porous solids: principles, methodology and applications. Academic Press; 2013.
[171] de Luna MD, Flores ED, Genuino DA, Futalan CM, Wan MW. Adsorption of Eriochrome Black T (EBT) dye using activated carbon prepared from waste rice hulls—Optimization, isotherm and kinetic studies. J Taiwan Inst Chem Eng 2013;44(4):646–53.
[172] Mashayekh-Salehi A, Moussavi G. Removal of acetaminophen from the contaminated water using adsorption onto carbon activated with NH4Cl. Desalination Water Treat 2016;57(27):12861–73.
[173] Kalpana D, Cho SH, Lee SB, Lee YS, Misra R, Renganathan NG. Recycled waste paper—a new source of raw material for electric double-layer capacitors. J Power Sources 2009;190(2):587–91.
[174] Wei L, Sevilla M, Fuertes AB, Mokaya R, Yushin G. Hydrothermal carbonization of abundant renewable natural organic chemicals for high-performance supercapacitor electrodes. Adv Energy Mater 2011;1(3):356–61.
[175] Bhattacharjya D, Yu J. Activated carbon made from cow dung as electrode material for electrochemical double layer capacitor. J Power Sources 2014;262:224–31.

[176] Teo EY, Muniandy L, Ng EP, Adam F, Mohamed AR, Jose R, et al. High surface area activated carbon from rice husk as a high performance supercapacitor electrode. Electrochim Acta 2016;192:110−19.
[177] Ariyadejwanich P, Tanthapanichakoon W, Nakagawa K, Mukai SR, Tamon H. Preparation and characterization of mesoporous activated carbon from waste tires. Carbon 2003;41(1):157−64.
[178] Rajesh M, Sau M, Malhotra RK, Sharma DK. Synthesis and characterization of Ni-Mo catalyst using pea pod (Pisum sativum L) as carbon support and its hydrotreating potential for gas oil, Jatropha oil, and their blends. Pet Sci Technol 2016;34(4):394−400.

CHAPTER 3

Biomass gasification integrated with Fischer–Tropsch reactor: techno-economic approach

3.1 Introduction

Today there is a rise in the rate of energy consumption worldwide due to the fluctuations and improvements in economic growth, mainly in developing nations, such as China and India, which resulted in high depletion of fossil fuel resulting in a shortage of fuel reserves. However, many problems are also caused due to the poisonous gases released from the combustion engines leading to various issues, such as air pollution, general public issues, and also issues related to global warming. There are many contributors to this CO_2 emission from various sectors, mainly from the vehicle sector, which uses a large quantity of energy, such as petrol, diesel, and other gasoline products in a liquid state obtained from fossil fuel. Hence, an extensive study was conducted to study liquid gasoline fuel production from renewable resources. It is seen that the total energy consumed across the globe was nearly 28% by the transport sector [1]. The vehicle sector consumed over 35% of total energy from the total energy consumption in Thailand. It is mandatory to make use of 25% of total energy consumed must be made from renewable energy by the year 2021 by the ministry of energy, due to which diesel is now getting a replacement with new energy products at a range of 25 million liters per day 2.

In the energy production industry, biomass is actively used as a significant energy source as it is eco-friendly and adds a benefit by being CO_2 neutral. The ratio of energy resources available in India is depicted in Fig. 3.1. Thailand is recognized as an agricultural nation since it produces many farm residues and farming products. As per the records, it is seen that in 2010 sugarcane was grown in a high quantity, followed by rice. The residues of rice are known as rice straw that was obtained highest among all other biomass residues [2].

From 31.5 million tons of rice, the rice straw obtained was nearly 25.6 million tons. It is also seen that rice straw is received on a large scale, almost 90% as agricultural waste, and is also found been burnt in the paddy fields once harvesting is done, creating air pollution and many breathing issues. Rice straw is made of hydrocarbons. Hence, it is beneficial to convert the waste rice straw into an energy source as it is

A Thermo-Economic Approach to Energy From Waste
DOI: https://doi.org/10.1016/B978-0-12-824357-2.00001-2

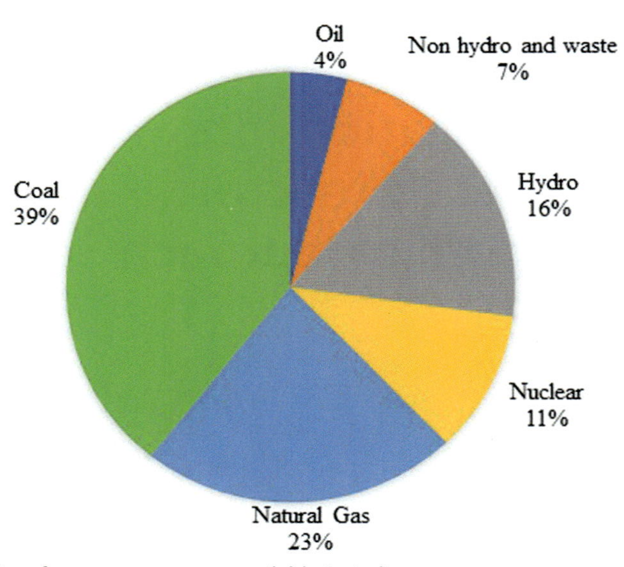

Figure 3.1 The ratio of energy resources available in India.

proven advantageous by safe disposal and energy recovery from rice industry waste. Along with the effects on the environment and the adoption of the latest substituted energy resource for internal production of energy. For the conversion of eco-friendly liquid fuels from biomass, either the gasification pertaining to biomass as well as that of the Fischer–Tropsch (FT) process or the biomass to liquid process has been recognized as the most capable method [3]. The ultraclean liquid fuels obtained from the FT process have no sulfur content and have a significantly less quantity of aromatic substances, leading to very fewer emission levels in the utilized combustion systems compared to other systems with fuels that are in the liquid phase formed from crude distillation. The ratio of renewable energy resources existing worldwide is shown in Fig. 3.2. This technology shall be done in the current structure as well as in the vehicle technologies too.

Thus the rise in pollutant gas emissions and the depletion of fossil fuels have made the BG-FT process pay more attention to technical and economic factors. However, this technology is still in the phase of research [4]. It is proven that technically BG-FT process had an operational period of 500 hours with stable conditions when it was tried for several runs. However, the researcher stated that the highest thermal efficiency obtained overall was 51%, along with 40% of the gasification and 75% of the FT process. In comparison, the BG-FT pilot scale showed 36.92% of the thermal process efficiency situated at the National Science and Technology Development Agency in the Thai city of Pathum Thani [5]. It is being observed that the highest exergy efficiency obtained from the exergy analysis done was nearly 36.4%. However, Tijmensen et al.

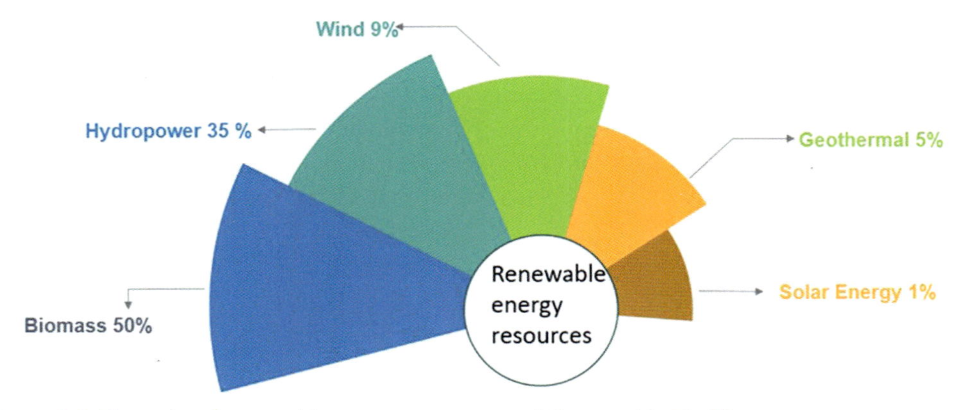

Figure 3.2 The ratio of renewable energy resources existing worldwide [8].

[6] stated that the efficiency of the overall lower heating value was found to be between the values of 33%–40% and 42%–50% in the gasification systems based on atmospheric and gasification systems, which are pressurized of the BG-FT process, respectively. It was also reported that the cost of production for the two concepts could not be compared with the current diesel cost as it creates a lot of difference. But the investigation report by the researcher stated that the BG-FT process performance in terms of techno-economic, which included the impact of variations in different kinds of the gasifying agent, which are air or enriched air and O_2, along with other factors like gasifying pressure, a configuration of the plant and cost invested and efficiency of current consumed by the plant, leading to the price of the diesel from the FT process [7]. Elkadeem et al. [89] showed a report of an economic analysis that was affected by various costs related to the configuration of the plant, which includes the investment cost on the plant, cost of operation on the plant, maintenance on the plant, devaluation, and funding charges on the price of the current consumed along with the liquid fuel synthesized. It was shown that the configuration of the plant is the main factor that decides the cost of both the products. However, it was also seen that even if the size of the plant was raised, the investment cost was very low. Several studies showed that the efficiency of the process of BG-FT could be improved by creating a proper integration of heat and combined heat and power system, along with enhanced feasibility in the economic terms by employing the configured system which uses bio-oil with complete conversion configuration [8]. These pathways, which are used for the production of the liquid transportation fuel and also for the production of electricity using the process of BG-FT as well as that of the Integrated Gasification Combined Cycle (where the feedstock is burned along with carbonaceous material), were examined with respect to the yield of total product obtained, which are electricity and liquid fuels along with emissions of carbon dioxide, and the cost of complete production [9].

However, previous research has conducted the study of the performances of the BG-FT process concerning technical and financial factors for creating a novel technique that offers liquid fuel for transportation that can be used for the replacement of current liquid fuel from crude distillation. However, there were many restrictions on the analyses. The obtained FT gas, which has syngas in an unreacted state (CO and H_2), can be reused to the upstream process like that of the gasifier or the FT reactor for increasing the yield of the product. But, the by-product consisting of CO_2 can be utilized as a gasifying agent that provides many benefits like there is no input of energy mandatory for the process of vaporization, with the gaseous product synthesized at a varied choice of H_2/CO ratios. Additional volatile substances are obtained in the phase of devolatilization produced by the reaction, namely the Boudouard reaction [10].

The main objective of the current work is to conduct the techno-economic analysis, which compares the dual arrangements and positioning pertaining to the process of BG-FT that utilizes rice straw as the primary feedstock, which includes the once-through and also the recycling of the concept of long loop where many parts of the off gas corresponding to FT are used with the help of the BG-FT archetype constructed utilizing the Aspen Custom Modeler software. As such, this investigation was done regarding the technical aspects of their effects of variation in the FT off gas recycle proportion for every constant FT reactor volume on the syngas processor execution, the FT synthesis process, and the overall BG-FT process. With the help of increased net present value (NPV), an investigation of the economic analysis was done to study the viability of the BG-FT process using the FT off-gas recycling when linked to the once-through concept. The study of factors like the economic rate of the plant's product, the plant's existent viability, and rate of return on the NPV parameter was done [11].

The effect of the rise of greenhouse gas emissions (GHG) causing impacts on weather and the dependence of power on fossil fuels is an alerting issue worldwide. Many nations had appreciated these queries to find different ways for different energy matrices by raising the ratio between renewable energy resources and locally available energy resources. It is seen that bioenergy acts as a critical role in many situations worldwide that focuses on the drop independence on the fossil fuels for gaining the targets of low-temperature stabilization. Like if we take the varied future situations, which are defined as per the Shared Socioeconomic Pathways, there will be mandates for the specific goal-oriented energy crops for the energy production that can range between 5 and up to 20 billion tonnes per year by the year 2100, with regard to nearly 0.2–1.5 billion hectares of property pertaining to this energy-based bio-production [12].

For some nations, such as Brazil, impetuses and stimulants were given for expanding the utilization of ethanol, which is obtained from the juice of the sugarcane plant, thus typical first-generation (1G) ethyl alcohol (ethanol) as the substitute for conventional fuels from 1970 to induce a comeback and counter the major oil disaster. After that, Brazilians' production costs on ethanol have reduced uninterruptedly because of the developments in the yield in agriculture and the scale and rise in efficiency through the conversion process. Thus 1G ethyl alcohol has raised a major part in the fuel discussion matrix and expressively contributed to decreasing GHG emissions in the vehicle sector. But the major challenge for the energy division in Brazil is to continually rise in the making of the bioenergy intended for the projected long-term growth for the demands in the energy sector all over the globe and also for the climatic commitments that was vowed as part of the 21st edition of the famed Conference of the Parties. For this, the Brazilian government has started preparing the new National Policy on Biofuels, also termed as RenovaBio, since 2016 [13]. This policy was put into effect from 2020 and formed instruments that provided economic incentives to the varied options available on liquid biofuel. The incentives were considered depending on the distinct performance of the production elements of the biofuel, which is comparative to the potential to decrease the release of GHG when compared to the fossil counterparts.

This integration between the first-generation and second-generation techniques has expected to raise the efficiency of the overall process and ease some risks that are related to the execution of new technologies for advanced biofuel. Many co-location approaches took benefits of the infrastructure shared, the rise in the prospects concerning the production with respect to power integration, as well as the rise in enterprise chances along with a greater collection of the goods with various applications, such as aircraft turbine fuels and light- and heavy-duty transport fuels. This literature also states the various combinations of the available biomass choices and the thermochemical systems, the opportunities for the integration of the gasification tailed by the FT synthesis using the characteristic 1G sugarcane distilleries, which has not been fully completely talked about and discovered. There are some latest reports in the study with regard to this integration emphasizing that still many feedstock types are available that are also promising and also the process designs that can be assessed. With regard to this, the chances of testing the process replacements along with the help of EC and wood chips from the eucalyptus as an alternative or substitute for traditional sugarcane (CS) happen to be unique. EC pertains to a specific cane that has high fiber content and also has high potential yields compared to CS. Hence, the utilization of EC is possible as an alternative to CS as our aim is to convert the LCM into liquid fuels of an advanced nature. Another interesting biomass happens to be Eucalyptus with respect to CS and EC, simplifying the operation process throughout the nonmain season. The tree has a big establishment in the supply chain in Brazil, with large volumes being produced, with major yields, and also with competitive costs. There are many possibilities of cultivating Eucalyptus marginal areas, which are not used for agriculture and which can be used to alleviate any extra external forces in play concerning using the land for bioenergy purposes [14].

3.2 Surplus biomass available in India

See Table 3.1.

3.2.1 Conflicting applications for crop residue biomass

There is a lot of difference in the fraction of surplus residue amongst different states in India, as shown in Table 3.2. Gross residue throughout the nation was found to be 34%, where Arunachal Pradesh had the minimum availability of 21% and Haryana and Punjab had the maximum availability of 48%. Uttar Pradesh had the highest residual amount in the entire nation with a percentage of about 33, similar to the gross value yet still lesser than several national states.

Table 3.1 Gross and surplus biomass available in India [15].

Crop group	Crop	Surplus potential (MT)	Gross potential (MT)
Horticulture	Arecanut	0.5	1.5
	Coconut	9.7	18.0
	Banana	12.3	41.9
Sugarcane	Sugarcane	55.7	110.6
Pulses	Lentil	0.3	1.7
	Gram	1.6	6.4
	Gaur	1.8	2.6
	Tur (arhar)	1.4	7.2
Oilseeds	Sunflower	0.6	3.8
	Groundnut	3.0	17.0
	Soybean	4.6	13.5
	Safflower	0.5	0.6
	Niger	0.0	0.1
	Linseed	0.0	0.3
	Sesame	0.1	0.8
	Mustard and rapeseed	4.9	12.7
Cereals	Jowar	3.5	17.6
	Ragi	0.3	2.7
	Small millet	0.1	0.6
	Barley	0.2	1.6
	Bajra	5.1	24.3
	Maize	9.0	35.8
	Wheat	28.4	131.1
	Rice	43.5	154.0
Others	Cotton	46.9	75.9
	Jute	0.4	3.9
Total (MT)		234.5	686.0

Table 3.2 Surplus of biomass state-wise (% of gross) [15].

State	Pulses	Horticulture	Sugarcane	Oilseeds	Cereals	Others	State avg.
Andhra Pradesh	23	44	40	26	29	38	33
Arunachal Pradesh	22	25	33	11	27	10	21
Assam	47	45	40	40	25	48	41
Bihar	23	20	33	32	27	10	24
Chhattisgarh	47	25	40	20	29	38	33
Goa	Not Available	39	38	NA	26	70	43
Gujarat	53	47	40	25	30	NA	39
Haryana	40	NA	40	37	34	90	48
Himachal Pradesh	35	25	38	17	35	48	33
Jammu & Kashmir	NA	NA	40	17	29	NA	29
Jharkhand	47	NA	40	33	26	10	31
Karnataka	47	32	38	31	30	33	35
Kerala	35	38	68	20	30	10	32
Madhya Pradesh	43	25	40	23	28	70	38
Maharashtra	47	43	33	33	28	40	37
Manipur	NA	25	40	21	28	NA	29
Meghalaya	35	20	40	15	26	30	28
Mizoram	35	32	40	18	29	48	34
Nagaland	35	43	40	16	27	38	33
Orissa	40	52	33	29	29	10	32
Punjab	47	NA	40	30	34	90	48
Rajasthan	23	NA	40	18	29	10	24
Sikkim	NA	NA	NA	22	28	NA	25
Tamil Nadu	30	48	40	19	33	29	33
Tripura	35	45	40	21	34	38	37
Uttar Pradesh	25	25	38	23	37	48	33
Uttarakhand	55	NA	38	53	32	NA	43
West Bengal	29	50	43	20	24	30	29
National avg.	38	42	39	30	29	38	34

3.2.2 Biomass

The general classification of biomass can be done as biomass from wood and non-woody biomass. Biomass from wood consists of output products gained from the trees sector, forest, and woodland. Nonwoody biomass consists of the farm crops, residues from the agriculture and forest section, animal waste, and tertiary waste, herbaceous products. It is seen that biofuel can only complete fossil fuel as they have similar conversion methods. But, many associated problems to the essential properties of lignocellulose biomass should be fixed [16].

3.2.3 Challenges in biomass utilization

There are benefits and drawbacks of each biomass, like other energy sources. The heterogeneous nature of biomass is considered to be the biggest drawback. Biomass feedstock can be classified into characteristics, such as physical, chemical, and morphological characteristics. Compared to fossil fuel, biomass has a very low density of energy and more water content in its unprocessed form. To produce the same amount of energy, more biomass is required when it is compared with fossil fuel. Biomass-based plants are generally bulky and can be easily affected by the attack of fungus and natural decomposition [17]. The susceptibility of biomass to degrade laterally with the distributed regions of many sources of feedstock are critically decreased by the attractiveness of biomass. The existence of wastes, lignocellulose content, spread distribution in ash content, and the influences, such as variation in climate, constraints of the location, and even changes in the cultivation learning, can lead to changes in the process. All these lead to more economical preparation of feedstock, handling of the biomass, and transportation of the biomass [18].

3.2.4 Biomass to energy conversion processes

The biofuel is of three types, mainly agro fuels, municipal by-products, and wood fuels, depending on the biomass utilized. Fig. 3.3 shows the manufacturing choice for preparing the appropriate energy carriers like liquid fuel or solid fuel obtained from biomass and bio-gas [19]. The methods can be bifurcated into three types, mainly classified as conversion by biochemical method, conversion by mechanical method, and conversion by a thermochemical method. Synthesized biofuel can be grouped into three main categories, which are agro fuel, municipal by-products and wood fuels.

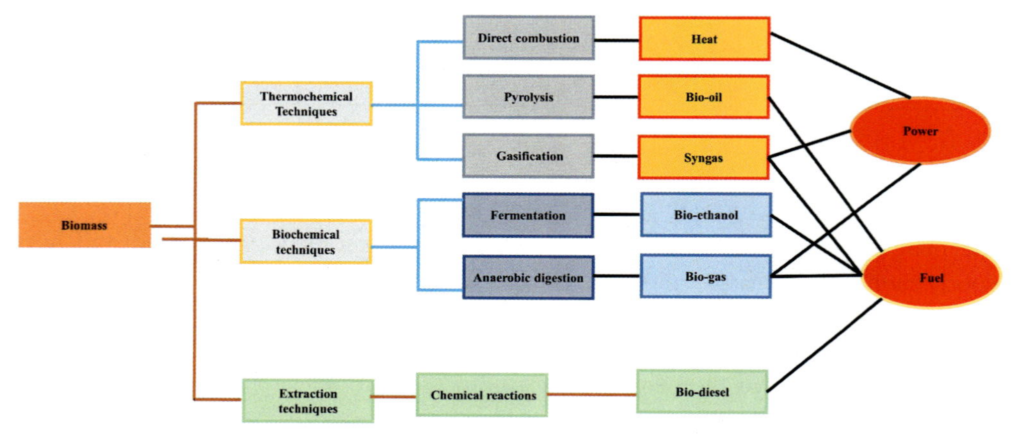

Figure 3.3 Techniques involved in the generation of energy from biomass.

Conversion by biochemical method makes use of the biological matters and the catalyst for converting the biomass into bioethanol, biogas, and biodiesel, which are convectional fuels used. Mechanical extraction from century-old technology is also considered as an option for gaining oil from a plant by physically rolling the seeds and crushing the seeds, kernel and fruits [20]. Heat and chemical catalyst is mainly required for the thermochemical processing for producing the useful secondary energy. Since this method has given rise to efficiencies, a rise in versatility, and fuel feedstock, it is considered the best alternative for converting biomass into energy. Biological conversion of biomass is a slower process when it is compared with thermochemical conversion of biomass. Benefits, such as compact emissions, energy efficiency, and thermal management improvement, as well as capabilities of producing hydrogen and other similar fuels, are offered by the gasification technology [21]. It is seen that the gasification process of biomass and coal is mostly the same as the yield obtained from thermal decomposition has the same products pertaining to a gas nature, which gives substantial leeway and leverage in price, processing as well as the range of secondary energy. Unfortunately, there exist drawbacks connected to this particular biomass, such as configurations of ashes possessing greater reactivity in the biomaterial compared to coal, which happens to be some disadvantages that need to be clarified and resolved. To resolve the deficiencies related to the undesirable features of raw biomass, pretreatment is done, which is promising a solution to raise the process efficiency before the chief energy alteration step takes place. Torrefaction, which is a pretreatment method that needs low temperature for the treatment, has been shown to have high efficiency for thermochemical processing. It shall be considered the primary pretreatment method as given in this paper [22].

3.3 Pretreatment of biomass

Biorefinery is documented as an effective technique for the world to do a practice to wean itself off limited hydrocarbons by using renewable biomass feedstocks to create a sustainable society. The most abundant natural substance on the planet is cellulose, which is similar to starch and contains glucose. But cellulose is such a material whose structure is much more difficult to deconstruct than sugar or starch, which has been used as a main bio feed in several refineries. This leads to the intimate association's intrinsic and close polymeric alignment with other lignocellulosic parts: hemicellulose and lignin. Even though many significant improvements have been made to improve the enzymes' ability to hydrolyze cellulose material, some pretreatment is still needed to ensure the recovery of the related lignin hemicellulose components in a better form and to enhance the enzymes' affability to the cellulose. Several lignocellulosic materials can behave as the feedstock for a biorefinery. The nature of the substrate is highly required so that it can be strongly influenced by the component recovery and cellulose accessibility enhancement by the type of required pretreatment, which includes lignocellulosic materials that possess comparably more flexibility to the micro/macro algae, which is being proposed as future feedstocks. In addition to the biorefinery approach, a thermorefinery strategy for converting biomass to various products has been posited, and this study has also been encased in this special issue. In the special case that has been profiled for many recent biomass pretreatment research and development actions in the biological and thermorefinery zones, improvements in technology developments, industrialization enhancement, and commercialization of biofuels and bio- and thermo-refinery processes have been highlighted [23]. This special issue has a total of 53 selected, peer-reviewed papers, which are grouped into four main themes which are: (1) pretreatment of lignocellulosic materials, (2) pretreatment of algal biomass, (3) pretreatment as it relates to a thermorefinery approach, and (4) others.

3.3.1 Torrefaction

Torrefaction is a thermolysis process that focuses the feedstock to thermal treatment at a low temperature of 200°C–300°C in the nonappearance of oxygen. Slow pyrolysis, roasting, thermal pretreatment, and mild pyrolysis are the terms that are commonly related to the definition of torrefaction for the uses. Previous studies on torrefaction were based on wood, such as wood chips and sawdust. But recently, many studies have shown the adoption of farm crops and residues from an agro-forest section. Table 3.1 summarizes the fuel-based characteristics pertaining to disparate biomasses. Even if many biomass resources were studied, the same product characteristics could be obtained through torrefaction, such as the enhanced value of energy, enhanced hydrophobicity, and friability, which is highly preferred. Analysis of the physical and chemical properties of the biomass prior and after torrefaction was done for obtaining the following: (1) yield, (2) energy content, (3) elemental composition, (4) change in major components, (5) hydrophobicity, and (6) ease of comminution [24].

3.3.1.1 Changes pertaining to that structure

When this process occurs, three main polymers are obtained when the biomass undergoes the physical and chemical variations: hemicellulose, cellulose, and lignin. These physiochemical variations rise to an idea to get the best condition for reaching the expected quality of the product. Tumuluru et al. [25] separated the torrefaction of these biomasses into three sectors at 50°C–150°C, known as the nonreactive zone, at 150°C–200°C known as reactive drying zone, and at 200°C–300°C, known as destructive drying. While heating from the temperature of 50°C to the temperature of 150°C, the water content disappears, and there is no account of chemical change. The shrinking biomass is obtained as the product of this process. While approaching the termination of this temperature between 120°C and 150°C, the lignin begins to get soften. Breakage of the hydrogen and carbon can be seen in the temperature range between 150°C and 200°C of the reactive drying zone. This course gives rise to compounds, such as fatty acids, saturated fatty acids, sterols, unsaturated fatty acids, and terpenes, which are also termed lipophilic compounds. Other factors at this temperature range are the deformation of the structure and the depolymerization process of the hemicellulose. The depolymerization process of the hemicellulose starts to short and condense the solid polymers. The rise in the temperature of 200°C–300°C leads to carbonization and devolatilization. In this temperature between this range, the whole deprivation of hemicellulose and half-finished deprivation of cellulose and lignin occurs. Yang et al. reinforced this statement by finding that lignin pyrolysis will appear only in the range of temperature between 150°C and 900°C. In comparison, hemicellulose pyrolysis happened to be observed in between the temperature of 220°C–315°C and cellulose between the region pertaining to the temperatures of 315°C–400°C [26].

With the help of many tools, many structural deformations are observed in the biomass at the time of torrefaction. With Fourier-transform infrared spectroscopy (FTIR), variations in functional groups are observed when the torrefaction process occurs. The factor that differentiates the raw and biomass undergone torrefaction process is reduced in the absorbance linked with O–H stretching vibration [27]. This results in a hydrophobic characteristic and decreased intensity of the absorption of hydroxyl that shows a decrement in the hydroxyl group [28]. The range of several external dominant peaks is seen in the spectra between 2800 and 3000/cm. After torrefaction, it is seen that the intensity of the dominant peaks reduces gradually [29]. But it was clearly stated that these outcomes are in sequence with the results, and hence, there should be a reduction in the peaks intensity and rise in torrefaction temperatures. Kopczyński et al. [30] investigated that the C–H stretching is clearly shown by the absorption in that range. More studies from Kasim et al. [31] showed that the CH_2 from hydrocarbons or carbonyl of the C–H stretching in the biomass are linked to these peaks. To raise the temperature of torrefaction, weakening of these peaks can be done, which is confirmed with the deprivation of the carbonyl group at a range between 1516 and 1560/cm or by changing the peak between 1790 and 1650/cm. Prado et al. [29] showed that the range's absorption characteristics show the stretching of C = O of the carbonyl group. CO_2 is released when the deprivation of the carbonyl group is done with the help of the decarboxylation mechanism.

To study the changes that occur at the molecular level at the time of biomass torrefaction, we can use different instruments: nuclear magnetic resonance spectroscopy. Around 245°C, it is seen that the degradation of the components begins of carbonyl carbons of acetyl groups and methyl carbons found in acetyl groups in hemicelluloses, which are seen at 21 and 172 ppm [32]. This deprivation of C4 and C6 in messed cellulose—during low (the temperature at 270°C) and mild torrefaction (the temperature at 300°C) is shown with the help of the signal declination [33]. Etherified guaiacyl is formed, as only a single position of syringyl (C3 or C5) in etherified β-O-4 structure has to undergo demethoxylation, which is seen at a temperature of more than 200°C that shows the lignin degradation [31].

3.3.1.2 Physiochemical properties

This process leads to transpositions in the biomass characteristics, such as fixed carbon, moisture, hydrophobicity, ash, grindability, and density and volatile content. Below subsections discuss the variations in these properties because of the torrefaction process.

3.3.1.3 Moisture content

Another key factor associated with fuel is the moisture content, as it has more moisture content, leading to more energy loss during combustion. But it is seen that the biomass with less moisture has more stability while storing, which has a little risk of biological deterioration. Also, hydrophobic solids can be transported at a very less cost [34]. When the biomass is placed at a surrounding with high moisture content, limitations are applied to the drying process as it is unable to stop the biomass from additional moisture adsorption [35]. Hydrophilic properties have resulted as biomass are highly found in the hydroxyl group [36]. Absorbed moistness is linked to the hydroxyl groups that happened to be present in the wall of the cell and is components of the lignocellulosic biomass [37]. Lowering the equilibrium moisture content can be founded by breaking down the hydroxyl groups with the help of thermal pretreatments, such as torrefaction. Andersson et al. [37] used dried pine fragments and yellow pine logging residue fragments as torrefaction feedstock, with the initial moisture content of 6.69% wet basis (wb) and 7.94% wb, respectively. After 30 minutes of torrefaction at 225°C, the moisture content of pine chip wb and yellow pine logging residue chip was indeed 3.30% wb and 3.11% wb, respectively. Jędrzejczyk et al. [38] used energy sorghum and sweet sorghum bagasse as feedstocks for the torrefaction process, with preliminary moisture contents of 7.83% and 9.29%, respectively. Once torrefaction is done at 225°C, after 30 minutes, it is seen that the moistness content was decreased to 2.05% for energy sorghum bagasse and 5.51% for sweet sorghum bagasse.

3.3.2 Types of pretreatment

3.3.2.1 Physical pretreatment

These methods of physical pretreatment in this pretreatment process include the usage of mechanical methods or irradiation methods and ignoring the usage of exterior foundations, such as chemicals and microbes in this pretreatment development. Initially, this method of physical pretreatment was used for changing the surface area, particles size, crystallinity index, and degree of polymerization of the biomass [38]. This method of physical pretreatment usually consisted of the proper steps with very low ecological issues, but this technique requires high power and energy consumption leading to a costly production rate. Thus ideas on the physical properties of biomass and the methods involved are very important for the choice of the physical dispensation of the biomass [38].

3.3.2.2 Mechanical methods

The most used methods used are mechanical pretreatment, which includes milling, grinding, and chipping. The most important stage in the pretreatment course is the size reduction technique. For modification of the assembly of biomass and for breaking the crystalline structure of the cellulose, it is very important to undergo this process that may later cause the surface area to rise for the additional process of the biomass. Moreover, it was specified that with the help of pretreatment, it is possible to decrease the size of the biomass can overwhelm the restrictions on the transfer of mass and heat [39]. Drop in the size may be used in many of the investigations of hydrolysis with the most well-organized method to enhance the enzymatic reaction of the lignocellulosic biomass.

3.3.2.3 Biological pretreatment

Pretreatment with the help of the organic way is taken as an ecologically and active course with little need for energy when related to physical and chemical approaches. Treatments, such as enzymatic treatment, microbial treatment, and fungal treatment, are the most widely known biological treatments [40].

3.3.2.4 Enzymatic pretreatment

Treatment with the help of enzymes includes the use of unprocessed, pure, or partly processed enzymes, which are gained from enzymes of hydrolytic enzymes and ligninolytic enzymes. It was described that the amount of lignin removed from biomass of lignocellulosic with the help of the enzymes from ligninolytic is nearly the same as that of the lignin elimination utilizing the pretreatment with fungus [41]. As of now, the efficiency of the complete enzymatic process in the process of lignin elimination is shown as very less cost when related to the old-style thermochemical methods. If we consider the current researches for the generation of bioethanol with the utilization of sugarcane bagasse pretreatment, along with alkaline with the help of NaOH and also with a ligninolytic enzyme, which is taken out from *Pleurotus ostreatus* IBL-02, pretreatment of the biomass was done independently. These studies of pretreatment reported that 48.7% and 33.6% of lignin elimination were gained from methods pertaining to alkaline and enzymatic modes, respectively. It is seen that the discovery of the enzymatic pretreatment methods of the bagasse gained has a high yield of cellulose hydrolysis, about 72.9%, and at the same time, alkali pretreated biomass gained less yield of 69.2% in the downstream progression of cellulose hydrolysis. These readings are highly influenced by the changes of cellulose fibers due to the enzyme removed that has both enzymes of ligninolytic and cellulolytic. These enzymes lead to the changes in the cellulose assembly and a rise in the rate of hydrolysis [42]. There are many enzyme-related aspects that affect the hydrolysis progression that consists of the inhibition of product, concentration of the enzyme, unproductive lignin binding, thermal activation failure, and adsorption of the enzyme [23]. The cellulose assembly acts as a crucial play in the enzymatic rate of hydrolysis. Degree of polymerization, arrangement of cellulose crystallization, approachability to the surface area, size of particles, and the presence of hemicelluloses and lignin are the basic benefits of cellulose that are affected by the hydrolysis process [43].

To decrease the thick cell walls, the utilization of hydrolytic enzymes is proven to be capable in the hydrolysis process of the bio-polymer for the microalgal biomass pretreatment [44]. Many products, such as CH_4, bio-H_2 gas, biodiesel, and bioethanol, were used to conduct many types of research of pretreatment of enzyme on microalgae [45]. Single or several combined enzymes can be utilized for the conduction of the use of an enzyme for the pretreatment method. Clean enzymes or the combinations of the enzymes, which contain unprocessed crude segregated from the organisms, such as bacteria and fungi, can be used for the enzyme cocktails. Several studies showed that the sole enzyme pretreatment technique had given a greater product yield with respect to the biomass, which is not processed [44]. In addition, other research showed a capable way of the cocktail enzyme for the cell breaking of microalgae biopolymers was shown. The utilization of cocktail enzymes was shown as the most utilized enzymes than single enzymes because of its feasibility for attaining the enzyme isolating in excess of one hydrolytic microorganism than their cleaned enzymes [46].

3.3.2.5 Microbial and fungus prevention pretreatment

This mode makes use of microbes that has boosted extensive researches in the recent few years as it is more beneficial with low energy consumption, ecologically friendly, very cheap cost on production, and low inhibitor with respect to the pretreatment by chemical means [23]. Fungus with varied varieties, such as brown fungi, white fungi, and soft-rot fungi, was utilized for biomass pretreatment. They are mainly dealt with the deprivation of lignin, hemicelluloses, and an insignificant amount of cellulose [47]. White-rot fungi are considered to be most effective for lignin removal when compared with another fungus available as they have to distinguish ligninolytic process [48]. It was stated that some white-rot fungi, with the help of the oxidation process, are capable of taking out the complete lignin from the lignin molecules [48]. Also, the decay of lignin with the help of soft-rot fungi is taken as an unfinished and slow step [41]. Several studies stated that soft-rot fungi have reduced lignin that is found in the angiosperm wood and can provide assistance in the diminishing of the secondary cell wall. For breaking down the structures that are nonphenolic, the soft-rot fungi may require a component that has a very less quantity of lignin. The changes in cellulose and hemicelluloses and also the minor modification is occurred because of the basidiomycetes or also known as brown-rot-fungi, as it is accountable for this (Table 3.3) [49].

3.3.2.6 Other latest pretreatment

This utilization of the innovative methods of pretreatment helps us to decrease the intractable characteristics of the microalgal biomass and the lignocellulosic biomass for liberating the captured polysaccharides for the generation of chemicals which has great worthiness that are additional which has been an exploration expedition for more than two centuries. But discovery of more supportable pretreatment is added active and ecological results that are the recent attention for high usage in industrial. Gaurav et al. [51] have newly proposed a distinct principle that has many supportable pretreatment should bear by: (1) changes in cellulose contented for improving the hydrolysis of downstream enzyme; (2) decrease in wastage of hemicelluloses content and cellulose content; (3) deficiency of creation of undesirable inhibitors; (4) energy budgets to be low; (5) decrease in demand for added size drop of biomass previously to the reaction; (6) compact waste and external products; and (7) little usage of chemicals. Existing pretreatment means are yet not able to upcome with the manufacturing claim of appropriateness. Thus advances in chemistry, hydrothermal methods, and thermochemical strategies have been used to create a natural and extremely dependable replacement for lignocellulosic biomass used in industrial applications.

3.4 Kinetics of biomass gasification for syngas generation

The thermochemical conversion in which the fuel obtained from solid biomass is converted to burnable fuel in addition to some amount of O_2, which is less than what is needed for the stoichiometric combustion, is termed as biomass gasification or gasification process of biomass. The reactor used for conducting the gasification process is known as a gasifier and includes many reactions for gas forming, which leads to transformation from the carbonaceous materials to burnable gases and fuels in a liquid state for many uses, like in engines and turbines, uses in direct heating and in fuel cells. Even if it is a hopeful method for uses in the production of thermal and power, the process of biomass gasification has many technical restrictions, which prevent its commercialization. The syngas has been promising to be a substitute and best fuel in operations of engines for transportation and power production. It generates power at a cheap rate and very less danger to health by the emissions of gases and by maintaining the neatness of the atmosphere [52]. To gain the advantages of the syngas as fuel to the engine, many changes and needs must be made in the engine. The syngas should possess all properties as that of an ideal fuel to ensure the operation of the engine is

Table 3.3 Different kinds of drying techniques [50].

Dryer type	Heat source	Feed mode	Drying capacity	Drying cost	Remarks
Solar dryer	Sun	Batch	Low	High	It was seen that drying a Large amount of biomass and drying the biomass in a specified climate faced many difficulties
Rotary dryer	Waste heat from flue gas	Continuous	High	High	The dryer was capable of drying large-scale biomass, but some influence of moisture was observed in the drying rate of biomass feedstock
Rotary dryer for filament biomass	Hot air	Continuous	Low	High	The drying system has great emissions of matters and organic compounds that are volatile. They are also prone to fire hazards
Superheated steam dryer	Superheated steam	Continuous	High	High	The dryer was eco-friendly and was able to recover up to 80% of energy, but it failed to dry large materials
Cascade and fluidized-bed dryers	Steam	Continuous	High	High	Dryers were affected by corrosion and erosion, and heat recovery was arduous to obtain uniform particle size
Belt conveyor	Waste heat from flue gas	Continuous	High	High	Usage of waste heat from leftover waste gas was used to reduce capital costs but increased emissions
Perforated floor dryer	Cylindrical air heater	Batch	Low	Low	The large, invariable vertical gradient was seen in the moisture content in the drying bed
Thermal screw dryer	Hot air	Continuous	High	Medium	Solid-to-solid heat transfer was high in this type of dryer

effective. These factors consist of gas purity should of high calorific value with low tar content, less than ($<100\ mg/Nm^3$) and the nonpresence of harmful products like NH_4 and SO_2 [53]. However, the syngas properties change with respect to the end uses, like reforming of the gas and engine uses for the generation of electricity and also for transportation. But, the main need for syngas is the same as that for normal fuel. Many types of research have shown the relation of producer gas obtained via the process of gasification undertook by various researchers all over the globe in recent years [54,55]. As many reactions are involved in the gasification process and subsystems, the quality of syngas is influenced by the type of feedstock used, design of the reactor, operating conditions, and gas cooling and cleaning methods [56] (Table 3.4).

3.4.1 Gasification mechanism

3.4.1.1 Drying zone or bunker section

The most crucial role in the process of biomass gasification is done by many physical properties and chemical properties of the unprocessed biomass material. Several studies shows that the type of biomass used is greatly influenced by the quality of the by-products obtained from the gasification process. Several studies on the process of gasification with the help of much-varied biomass feedstock was investigated in different research articles and also briefed in this specific one. It was also shown that many processes like the pretreatment process had been carried out on the unprocessed biomass materials, such as briquetting, drying, and densification, to emphasize the quality of the by-product. But, it is showed that the process of drying biomass occurs inside the gasifier reactor on its own. There is a high quantity of water content that leads to high loss of energy and also drops the product quality. The presence of the moisture content quantity may generally change in between the range varying from the value of 5% to the value of 35% that totally influenced by the nature of biomass that is going to be transformed into steam at a temperature that nearly becomes 100°C. As there is a transfer of heat pertaining to the combustion zone, it affects the drying process of the biomass that occurs in the section of the bunker. It is also seen that the fuel has no effect on the thermal decomposition of the volatile substance at the time of drying because of low temperature, as shown below [22,59].

3.4.1.2 Pyrolysis or thermal decomposition zone

The process when fuels obtained from biomass are treated in the absence of air or O_2 where thermal decomposition takes place is known as pyrolysis. This step gives a rose to liquid tar, solid charcoal, and gas product where the proportion of the by-product is influenced by the nature of fuel obtained from biomass, which is utilized here along with the operating conditions of pyrolysis.

When pyrolysis takes place, there is a combination of the drying process and reduction process of the molecular weight of the substituents that occurs at the same time, and the

Table 3.4 Summary of significant factors affecting gasifier performance.

Factors	Significance	Ref.
Equivalence ratio	The equivalence ratios are highly affected by the process of gasification and also by the by-products formed along with the produced tar content provided with high heat. The higher ratios of this also lead to the huge reduction in heating values pertaining to producer gas. High steam or ratio of biomass leads to a greater yield of H_2 and also cracking of the tar content, which is caused due to water shift reactions. The equivalence ratio must be between 0.2 and 0.4 to gain better gasification efficiency	[57]
Moisture content	Several studies have investigated the equilibrium moisture content that acts as a strong factor for the relative humidity and a weak factor for the atmospheric temperature. The moisture content is inversely proportional to the rate of the biomass consumed as it is observed that greater the moisture content, there is a drop in the rate of the biomass consumed, leading to decrement in the pyrolysis of biomass. However, the performance of the gasifier and the quality of the by-products formed are highly influenced by this factor. It has been stated that in a downdraft type of gasifier, the moisture content has a greater value of nearly 40%	[58,59]
Superficial velocity	Process of slow pyrolysis results from decreasing in values of the superficial velocity that leads to a greater yield of char and also unburned tars. At the same time, it has been studied that the yield of char is dropped due to the high value of the superficial velocity, which is caused because of the process of fast pyrolysis. But, there is some reduction in the time required for gas residence resulting into a decrease in the efficiency of the tar cracking method. It has been studied that this velocity may vary in between the range of from 0.4 m/s and up to 0.6 m/s and is best for the internal combustion engine	[60,61]
Operating temperature	Several investigations have been made stating that the operating temperature of the high gasifiers, which has the capability for converting the high biomass carbon resulting from dropping in the tar produced and also led to a generation of more burnable gases. But many studies have proven that the concentration of H_2 is seen to be rising at the starting stage, and then there was a drop in the concentration with a rise in the temperature	[62]
Gasifying agents	Many investigations showed that the gasifier atmosphere is the most important parameter considered as a dependable factor to the by-products formed by the gasifier system. For the generation of syngas with a very low heating value, the air is preferred as the gasifying agent as it is able to dilute with the help of nitrogen and at the same time, the moisture if both steam and O_2 gains syngas that has a heating value of a moderate or medium nature. But when steam happens to be in mixed form with the air leads to the production of a high yield of H_2 that leads to a drop in the energy needs of a system	[58,60]
Residence time	Several studies have reported that the formation of tar and the composition of tar are highly affected by the residence time. It is also seen that a study showed, drop in the O_2 consisting of elements with added one and two aromatic ring compounds lead to a rise in the residence time. But when the elements have three and four rings, it is stated that there is a rise in the residence time. Other investigations show that a drop in the content of tar produced has been seen in the process because, in biomass gasification, augmentation of space-time takes place, which has dolomite bed	[63,64]

water content gets off when it goes below 200°C. The molecular weight reduction process of the biomass occurs when the given temperature rises to 300°C, and those substituents include amorphous cellulose that initiates the creation of carbonyl group radicals and carboxyl group radicals. At the time of the reduction process, CO and CO_2 are also produced. It is seen that tar, gaseous by-product, and char are produced by the degradation of the formed crystalline cellulose when the heat is increased above 300°C. Production of char, volatile gases, and tar are obtained when the hemicellulose undergoes degradation to form the soluble polymer. The formation of methanol, acetic acid, water, and acetone occurs when the lignin gets degraded from a temperature above the range from 300°C and up to 500°C. It is also investigated that carbon (char) and medium-size molecules (volatiles) are formed by the conversion process of cellulose, hemicellulose, and lignin, which are large biomass biopolymers. Thus it shows that the pyrolysis process of the biomass usually occurs in between the temperature range of from 125°C and up to 500°C, where the condensation process is also done for the production of tars from the hydrocarbons. Exothermic reactions are reactions that occur when a chemical reaction commences at a heat of 300°C, and endothermic reactions are reactions that occur when a chemical reaction when the heat is given is above 300°C. For the production of charcoal, no external energy is required for heating, as 300°C is enough for the production. In order to maximize the yield of the liquid or gaseous fuel, external heat supply is essentially the process of pyrolysis at high temperature, as shown below [21,65]:

$$\text{heat} + \text{feedstock(dry nature)} \rightarrow \text{char} + \text{volatiles} \tag{3.1}$$

3.4.1.3 Partial oxidation or combustion zone

It is seen that the exothermic type of chemical reaction takes place where the volatile materials contained by the biomass undergo oxidation and forms oxide leading to the formation of heat with the temperature peak value changing in between the range of 1100°C and up to 1500°C along with the by-products like gases of CO_2, H_2, H_2O, and CO during the oxidation period of time-frame. This is the most important step in the gasification process since it determines the quality and variety of by-products produced. Factors like the reactors' inbuilt pressure and temperature and the gasifying agent used for gasification, such as O_2, air, or steam, act as important parameters for obtaining high yield of the syngas. Several studies have shown that the steam acts as the best medium for gasification, which supports the process of gas reforming. But it is also studied that atmospheric air with a large quantity of NO_2 highly influences the heating properties of the output obtained from gasification. But it has also been shown that O_2 is also acted as a satisfying agent for the gasification process as the output syngas used for the generation of power is noted. At the same time, the drying process of biomass is supplied heat from the heat generated in the combustion zone undergoing exothermic reaction. This heat is also used in the pyrolysis process for getting the volatile matters and also to give heat for the reduction reactions. In such zones, fuels from solid carbonized along with O_2 in the atmosphere undergo heterogeneous reactions in this zone that leads to the generation of CO_2 and some heat. H_2 also mixes with O_2 to form H_2O vapors [21,65]:

$$C + O \rightarrow CO + 406 \text{ kJ/g mol (complete oxidation reaction)} \tag{3.2}$$

$$2H + O \rightarrow 2HO + 242 \text{ MJ/kg mol } 22 \tag{3.3}$$

3.4.1.4 Reduction zone

The process of gasification/pyrolysis generates utilizable as well as harmful products, such as gases as nitrogen oxide and sulfur oxide, and many contents of tar products are shown as the most obstructed materials that block usage of the syngas produced for the uses like production of power. Many types of research have proved that when operating conditions are set to be different, as shown in the next article, tar formed is seen as the important point by the researchers. Mechanism methods, self-modification methods, and thermal decomposition strategies are examples of tar removal techniques. The researchers have tested and analyzed the catalyst cracking and plasma processes. The usage efficiency of the biomass is highly affected when a large quantity of tar is found in the fuel gas and can cause severe problems, such as clogging at the points of fuel lines, fuel filters, and also in the engine—keeping fixed temperature conditions that are able to control the tar formation that is needed for thermal decomposition at the reduction zone. It has been investigated that a temperature range of around 1000°C at the reduction zone is needed for the drop in the tar contents. Also, many chemical reactions occur under reducing the air that gains transforming of the mixture of heat in gases and charcoal to form chemical energy of the syngas. The endothermic reaction takes place when this process of reduction process, leading to the formation of burnable products, such as carbon monoxide, hydrogen gas, and methane gases [21,65] (Tables 3.5 and 3.6).

$$C + CO \rightarrow 2\ CO - 172.6 \text{ kJ/g mol (Boudourd reaction)} \tag{3.4}$$

$$C + HO \rightarrow CO + H - 131.4 \text{ kJ/g mol (water gas reaction)} \tag{3.5}$$

$$CO + HO \rightarrow CO + H + 42.3 \text{ kJ/g mol (water gas shift reaction)} \tag{3.6}$$

$$C + 2H \rightarrow CH + 75 \text{ kJ/g mol (hydrogasification reaction)} \tag{3.7}$$

Table 3.5 Chemical reactions involved in the gasification process [61,62].

No.	Reaction	Type of reaction (kJ/mol)	
1	$C + CO_2 \leftrightarrow 2CO$	+ 172	Carbon reaction (Boudouard)
2	$C + H_2O \leftrightarrow CO + H_2$	+ 131	Carbon reaction (primary steam reforming)
3	$C + H_2O \leftrightarrow CO_2 + H_2$	+ 90	Carbon reaction (secondary steam reforming)
4	$C + 2H_2 \leftrightarrow CH_4$	−74.8	Hydrogasification
5	$C + 0.5O_2 \rightarrow CO$	−111	Oxidation reactions
6	$C + O_2 \rightarrow CO_2$	−394	
7	$CO + 0.5O_2 \rightarrow CO_2$	−284	
8	$CH_4 + 2O_2 \leftrightarrow CO_2 + H_2O$	−803	
9	$H_2 + 0.5O_2 \rightarrow H_2O$	−242	
10	$CO + H_2O \leftrightarrow CO_2 + H_2$	−41.2	Shift reaction
11	$2CO + 2H_2 \rightarrow CH_4 + CO_2$	−247	Methanization reactions
12	$CO + 3H_2 \leftrightarrow CH_4 + H2O$	−206	
15	$CO_2 + 4H_2 \rightarrow CH_4 + 2H_2O$	−165	
13	$CH_4 + H_2O \leftrightarrow CO + 3H_2$	+ 206	Steam reactions
14	$CH_4 + 0.5O_2 \rightarrow CO + 2H_2$	−36	

Table 3.6 Composition of producer gas in different biomass feedstocks [66].

S. no	Biomass	Producer gas composition (% by vol, dry basis)					Lower heating value (MJ/Nm^3)
		H_2	CO	CO_2	CH_4	N_2	
1	Olive pitts	21.81	23.91	7.94	0.21	46.12	5.91
	Hazelnut shell	20.29	25.79	8.28	0.18	45.47	5.94
2	Bamboo wood	21.91	23.14	9.60	0.21	45.15	5.82
3	Fir mill	21.97	25.64	8.59	0.21	43.60	6.15
4	Coconut shell	21.53	25.59	9.17	0.20	43.52	6.08
5	Softwood (av.)	21.16	24.63	8.75	0.19	45.27	5.91
6	Spruce wood	21.30	24.79	8.41	0.20	45.31	5.95
7	Hybrid poplar	20.76	23.44	9.18	0.19	46.44	5.71
8	Willow wood	21.19	24.33	9.22	0.20	45.06	5.88
9	Hybrid poplar	20.76	23.44	9.18	0.19	46.44	5.71
10	Pist. shells	21.94	23.93	8.88	0.21	45.04	5.93
11	Wood bark (av.)	21.03	24.98	8.35	0.19	45.45	5.94
12	Akhrot shell	21.14	25.12	9.35	0.19	44.19	5.97
13	Almond shells	20.40	22.92	9.58	0.18	46.92	5.59
14	Corn cob	21.04	25.49	9.79	0.19	43.49	6.00
15	Sugarcane bagasse-1	20.95	23.50	9.92	0.19	45.44	5.74
16	Casurina wood	21.91	23.26	9.70	0.21	44.92	5.84
17	Hardwood (av.)	21.26	22.78	9.75	0.19	46.01	5.69
18	Groudnut shell	18.64	21.65	10.31	0.15	49.26	5.19
19	Neem wood	21.97	23.14	9.83	0.21	44.86	5.83
20	Almond hulls	18.95	20.51	10.47	0.15	49.91	5.09
21	Olive husk	20.33	22.28	9.88	0.18	47.34	5.50
22	Ply wood	21.12	23.42	9.83	0.19	45.44	5.75
23	Mixed paper	18.54	18.56	10.39	0.15	52.37	4.79
24	Jujuba wood	21.59	22.86	10.20	0.20	45.16	5.74

Mechanism methods, self-modification methods, and thermal decomposition strategies are examples of tar removal techniques.

3.4.2 Syngas conditioning

The syngas generated at the time of gasification of biomass includes a combination of a variety of gases, such as CO, CH_4, N_2, H_2O, H_2, and CO_2, with other waste substances, such as tar contents, solid substances, ammonia, HCl, and other sulfur elements and species of alkali metals. These harmful waste gases produced cannot be utilized at the end. Hence, for gaining working and efficient operations of the applications of engines and production of the biofuel, it should be cooled at low temperature and must be kept in a good and clean condition.

Some parameters, such as the design of the gasifier, type of feedstock used, gasifying agent used for gasification, and other process factors, highly influence the composition of gas and the concentration of impurities. The syngas produced at the output of the gasifier should be in cool condition prior to the end-user. Else it will drop the efficiency of the process, especially in the uses at downstream gasifier. Many cooling cools the hot air up to specific atmospheric parameters defined by a certain limit. For other arrangements of gas cooling setup, the gas coolers that are used are natural and forced gas cooler where the heat in the hot air is transferred through the pipe length and no extra power is needed for its operation, and also constriction is very simple but huge in size. In addition, many electrical instruments like a fan are needed for transferring the heat via pipe and needs additional power for operating. At the same time, a water cooler is termed as the best method for gas cooling while other methods, such as scrubber and heat exchanger where tar contents are cooled and condensed, are made interact with the water which is being circulated. This leads to contamination of water leading to damage if cooler. At the same time, with the help of a water-cooled heat exchanger, it is possible to maintain the neatness of the circulated water, and the energy needed can also be maintained. The gas cleaning mainly consists of the removal of impurities, such as tar and solid particulates that stay over some range, as stated in the last section, and then the pragmatic due to the complexity in the process and restrictions to handle. But a limited quantity of such matters is taken into account for some uses in downdraft gasifier like in those of internal combustion (IC) engine turbines and also pertaining to fuel cells as shown in Table 3.7. Hence, processing the syngas happens to be the extremely crucial and main step for many uses of heat and electricity production. The cleaning methods available are physical process and chemical process for cleaning the syngas as shown below [63,67].

Table 3.7 Presence of pollutants in producer gas and used controlling mechanisms [64].

Pollutant	Source	Possible problems	Control mechanism and/or mitigation
Particulates	Ash, bed material	Erosion, agglomeration and fouling. Environmental pollution	Filtration, gas cleaning (scrubber)
Alkali metals (sodium, potassium in the ash)	Ash	Corrosion	Cooling, condensation, filtration, adsorption
Nitrogen compounds (NO_x, NH_3, HCN)	Reaction of nitrogen contained in air and feedstock	Corrosion, environmental pollution	Treatment with substances of basic character, use of pure oxygen in the process
Sulfur and chlorine compounds (HCl, H_2S)	Reaction of sulfur and chlorine contained in the feedstock	Cleaning, capture with $CaCO_3$, $MgCO_3$	Sulfur and chlorine compounds (HCl, H_2S)
Tar (complex hydrocarbon mixtures)	Low temperatures in the process, considerable amount of volatile in the feedstock	Corrosion, agglomerations and fouling. Health hazard	Removal, cracking

3.5 Gasification integrated with Fischer–Tropsch reactor

A model was designed that was integrated with the process of gasification of biomass, and FT-integrated process was made with the utilization of the Aspen Custom Modeler program. The components involved in the model are the biomass gasifier, synthesis gas cleaning and gas conditioning, and also the FT synthesis, as represented in Fig. 3.4 Initially, unprocessed syngas was generated from the biomass, which mostly constituted of hydrogen, carbon monoxide water, carbon dioxide, and some methane gas, along with some tar and char. At the same time, the unprocessed syngas was required to be properly cleaned so that its mixture meets the criteria of FT in sections of gas cleaning and conditioning. Transportation fuel in a liquid state is generated along with the FT off gas and water steam as the end product in the synthesis unit of FT [65]. Before it is recycled into the gasifier unit, the pressure of FT off gas is required to drop as that of gasifying pressure by moving it through the expansion turbine attached to a generator that leads to the production of some power. This FT off gas, which has low pressure, is segregated into two main sections: the first section is reused to the gasifying unit to enhance the yield of the product and the second section is flown through to flare or transferred to adjacent units as a fuel gas [65].

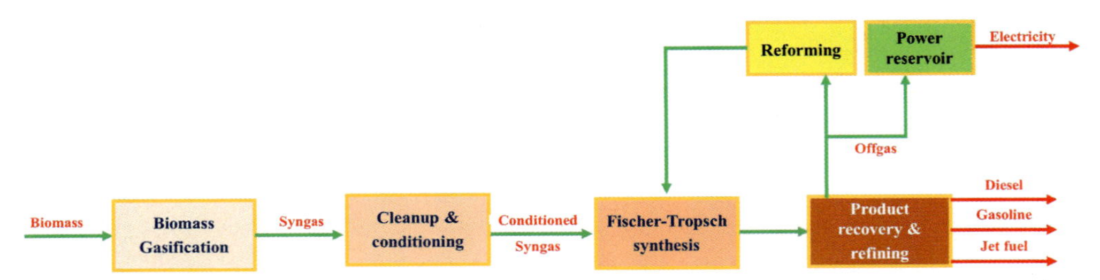

Figure 3.4 A schematic view of a biomass gasification plant for the manufacture of FT fuels and electricity [68].

3.5.1 Bioenergy potential calculations and estimation

Bioenergy potential is obtained from crop residue, and biomass is calculated using the following equations: [15]

$$E(j) = \sum_{i=1}^{n} \mathrm{CRs}(i,j) \times \mathrm{HV}(i,j) \tag{3.8}$$

where $E(j)$ is the bioenergy potential of n crops at the jth state (MJ); $\mathrm{CRs}(i,j)$ is the surplus residue potential of the ith crop at the jth state (tonnes); and $\mathrm{HV}(i,j)$ is the heating value of the ith crop at the jth state (MJ/tonne).

3.5.2 Fischer–Tropsch synthesis

In FT synthesis, syngas produced that is CO along with H_2 is changed catalytically into a spectrum of HC chains (Eqs. 3.9–3.12):

$$\mathrm{FTS}{:}n\mathrm{CO} + (2n+1)2 \rightarrow \mathrm{C}_n\mathrm{H}_2n + 2 + n\mathrm{H}_2\mathrm{O} \tag{3.9}$$

$$n\mathrm{CO} + 2n\mathrm{H}_2 \rightarrow \mathrm{C}_n\mathrm{H}_2n + n\mathrm{H}_2\mathrm{O} \tag{3.10}$$

$$n\mathrm{CO} + 2n\mathrm{H}_2 \rightarrow \mathrm{C}_n\mathrm{H}_2n + 2\mathrm{O} + (n-1)2\mathrm{O} \tag{3.11}$$

$$\mathrm{WGS}{:}n\mathrm{CO} + \mathrm{H}_2\mathrm{O} \rightarrow \mathrm{CO}_2 + \mathrm{H}_2 \tag{3.12}$$

Iron, nickel, ruthenium, and cobalt are considered as the four major metals for the Fischer-Tropsch synthesis (-FTS) process. Ruthenium is considered as an active metal, but this is not possible as it is economically high, and the abundance is also very low [68]. Nickel being very cheap has defects like it get plagued while coking and is taken to be a methanation catalyst [68]. Hence, CO and Fe are thus considered to be important metal in the industry for the process of FTS. The rate of FT catalyst that is iron-based is around the range of $10–$40 per unit pound; at the same time, the FT catalyst that is cobalt-based is rated at a range of $60–$100 per unit pound and is highly convenient to take to the marketplace as there is a rise in the needs in the field of battery systems and aviation. Each FT catalyst created till now possesses unique and different structures of yield [69]. Cobalt catalysts make a high quantity of H_2O. Hence, cleanup of water is needed. When using a Fe-based catalyst for WGS water gas shift, more carbon dioxide is produced and transported to a shift reactor. Many types of feedstock can be used in the catalyst based on iron with respect to cobalt catalyst, but, economically, it is the most demanded and the low-cost feedstock. Iron has other benefits like the removed CO catalyst that are required for the process of reclamation, but iron catalyst can be landfilled (only if the catalyst is under the nature standards for CO). When H_2/CO ratio near to 2 is to be synthesized. Cobalt may be mixed with gas to form a liquid [69].

3.5.3 Fischer–Tropsch catalysts

Iron-based catalyst mainly has additional structural and chemical promoters. Binders like silica are used in the Fe catalyst so that there are improvements in the rigidity of structure [70]. Alkali metals are usually electronic promoters that provide carburization, leading to a rise in the FTS activity and resulting in improvement in the selection of the high molecular weight of the hydrocarbons. Many studies have shown that the acceptance of iron along with many other metals, especially for the purpose of tuning, is described. Promoters like Cu provide benefits to the reduction of the FeO [71]. As cobalt is very costly, to raise the quantity of cobalt that comes in contact with the surface, cobalt molecules are connected with metal oxides, ceria, zeolites, manganese, many other numbers of materials, and carbon materials. Such elements are mainly inert and are utilized for promotion for a specified selectivity that will be brief in more detail afterward not like Fe, promoters used for Co will not rise the turnover frequency; else the metals aid rises with the % of removal of oxygen of cobalt oxides undergoing reduction to cobalt-based metal [72]. As it supports, methanation is influenced by the nature of the promoter. Moreover, unlike CO and ruthenium, the surfaces pertaining to that of a metallic nature are less supportive for FTS; many types of iron carbide show FTS working [73].

3.5.4 Fischer–Tropsch mechanism

It has been nearly 90 years since the foundation of FTS, even after that, there is much information with respect to the mechanism stays. Carbon monoxide or hydrogen gas is interactive for both active metals that are yet to be researched to find the specific pathway needed for the dissociation and their effect of kinetics on the chain growth of hydrocarbon. Carbon monoxide is interacted directly or with the help of hydrogen in a specified manner [68]. There are varied mechanistic members of the family, which has initiated to explain the process of initiation and also the steps of chain growth at the time of FTS that are the mechanism of the carbide and insertion of the CO. In carbide mechanisms, the initial proposing was done by FTS, and carbon monoxide actively leads to the generation of carbide and surface oxygen. Species of carbides are half hydrogenated to form the CH_x particles that serve as the chain growth monomer. Termination takes place when adding H_2 leads to the formation of olefins or by adding the CH_3 or H_2, resulting in kinds of paraffin [68].

In the mechanisms of CO insertion, carbon monoxide is directly put into the increasing CH chain before to C—O scission manner [74]. C—O scission is the main phase that helps the mechanistic scheme for the microkinetic models, and steady-state isotope, tracer studies, and transient kinetic investigations [75]. Isotopic kinetic tests were done by changing from the ratio of H_2/CO to the ratio of D2/CO and by keeping certain conditions that are T, P, space velocity, and H_2/CO for converting it into a steady state of the carbon monoxide. H_2-deuterium tests were done at the time of FTS, which has shown many rate ratios of kinetic but have only shown an effect of reverse kinetic isotope (IKIE) [76]. The conclusions of the test from isotopic surface separation show that chemisorption is irresponsible for the effect of H/D isotopic. Several studies of H/D separations show that a near-uniform that has H/D to be 0.97 combines with the H_2 isotopes residing on the surface of the active metal. The near-uniform partitioning conclusion, combined with the IKIE seen in H-D changing at that time, shows limitations to kinetic in the creation of H-bond that compiles with the founded values of IKIE [77].

3.5.5 Biofuel synthesis from Fischer—Tropsch reactor

To raise the efficiency to the max, it is necessary to optimize the configuration of reactor and process factors and also for gaining the expected product selectivity [78]. Here, the latest improvements in FTS reactors are shown. As FTS is greatly meant to be an exothermic reaction, managing the heat is very important. Many reaction temperatures result in a high yield of methane, deposition of carbon and deactivation of the catalyst. Slurry bubble column (SBCR), multitubular trickle-bed, circulating fluidized-bed, and fixed fluidized-bed reactors were used in the early stages of FTS (Fig. 3.5), and several review papers were written [79].

3.5.5.1 Slurry bubble column reactors

The improvement in the process of slurry in the process of FTS that runs in a very less temperature started from the term of 1938 introduced by the German researcher, and the method gained a generation range of 11.5 tons that can also be said as 75 bbl in a day of HC [80]. The gas that is synthesized in the SBCR comes through the reactor in the base section and combined properly with the liquid that consists of the catalyst [81]. When the surface of the catalyst reacts with the syngas departing from the top, a

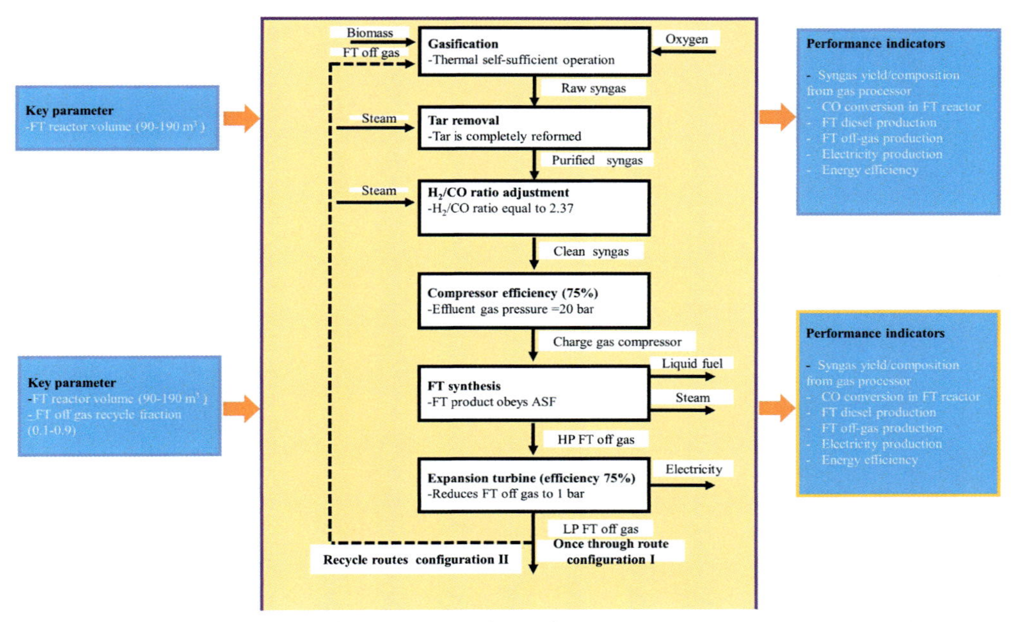

Figure 3.5 Different pathways of Fischer-Tropsch synthesis.

large amount of HC is collected from the base using a metal filter of average size. Cooling tubes extract the generation of the heat with the help of the reaction. The main problem with the slurry phase reactor is the bifurcation of the product from the catalyst. As there is some breakthrough of catalyst, which takes place, the activity stops. An issue with the typically used SBCRs is raised. There are many studies showing that the design of the reactor is only dependent on the conclusions of the reactor from the lab or from the readings of the pilot plant. Predictions are predicted depending on initial experiments on SBCRs at a very less scale that cannot be utilized at the industrial level due to changes made in reactor with respect to hydrodynamics. But the latest study by Van Steen et al. [82] shows that correct designs for the process of FTS in the SBCR are done for providing the scale-up for next usage. SBCR method has been used by many manufacturers and institutions as they are simple in operations with respect to the fixed-bed type of reactors. Deckwer et al. [83] investigated the influence of various factors on working of the reactor with the help of the models that has three phase; they stated a rise in the total change from 0.79 to the value of 0.97 by raising the temperature of the reaction from 258°C to the temperature of 90°C and also change in the pressure to the level of 3 MPa. Rising the ratio of the H_2/CO in the feed for a long time leads to dropping in the selectivity of C_21 and raises the selectivity of CH_4. Sasol shows the SBCR as given in Fig. 3.6. Many studies have been carried

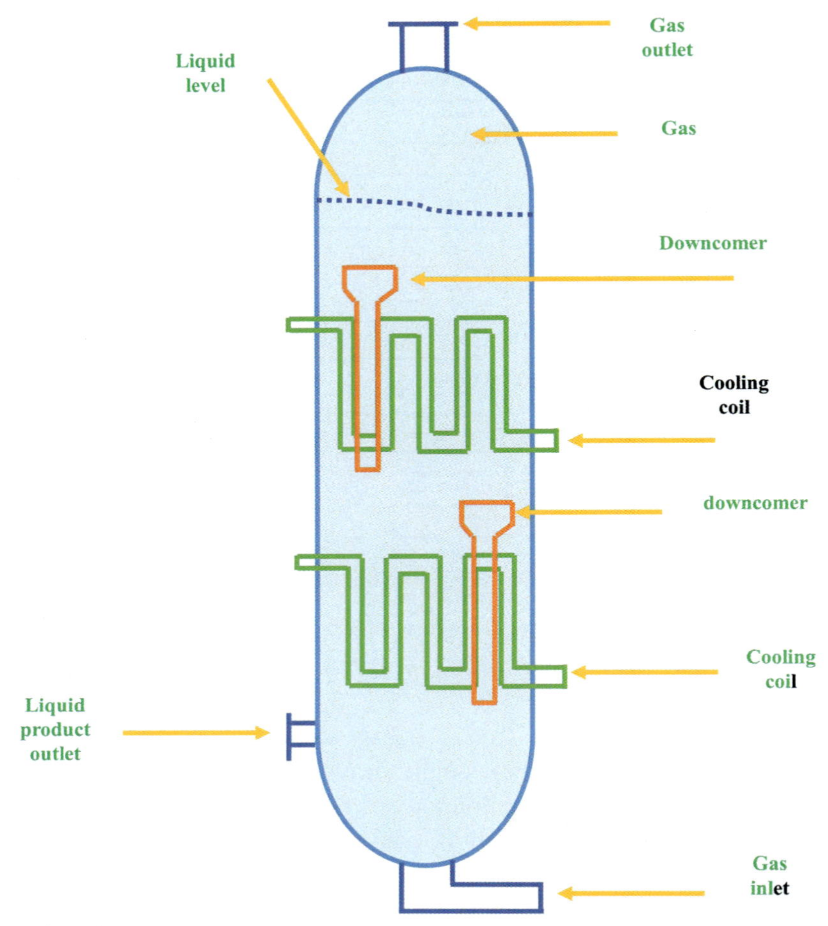

Figure 3.6 Slurry bubble column reactor.

out to determine the factors that influence the operation of a bubble reactor, as it allows the insights into churn-turbulent flow and also the models that are based on other flow patterns are also shown in the reality of operation of SBCR [68].When the surface of the catalyst reacts with the syngas departing from the top, a large amount of HC is collected from the base using a metal filter of average size.

3.6 Techno-economic analysis of Fischer–Tropsch reactor with biomass gasification

See Table 3.8.

Table 3.8 Techno-economic study of biomass gasification [84].

Author	Year	Significance	Ref.
Parajuli et al.	2014	A cost analysis model was developed to research the cost analysis of the development of syngas and noted that it is increased from $0.543–$0.043/Nm and increased the production capacity from 60 to 1800/Nm^3/h. It was also proposed that the cost of syngas production per unit will reduce if the production capacity is higher	[85]
Antonia et al.	2014	The techno-economic and combined heat and power biomass system integrated with gasification, combustion, and Stirling engines were taken into account with the net present value for the best technological solution. Gasification plant with combustion engine consuming about 4040 €/kWe with remuneration period of 4–5 years was not as reliable as it is with Stirling engine which requires high investment cost (7429 €/kWe)	[86]
Lundgren et al.	2014	The techno-economic analysis of NH_3 production via integrated biomass gasification in a pulp and paper mill by comparing the results with a standalone production case. The energy and economic performance were high compared with standalone production case. In this regard, a high NH_3 selling price was required for the economic feasibility of both cases	[87]
Muresan et al.	2013	The efficiency, output, gas composition, cost, and CO_2 capture rate of hydrogen generation utilizing biomass cofiring were all investigated and reported. It was noted that in coal gasification alone, the operational cost and the capital were the highest and reported to decrease as the quantity of the biomass increased. The operational cost was found to be high in coal gasification due to coal transport charges	[88]

3.7 Conclusion

The importance of biomass in the energy sector is now well known to us. The methods and techniques to extract energy from biomass and their optimization can surely support the energy sector and satisfy a part of the energy demand for people. Gasification is one of the important processes to produce syngas, which has a high H_2 to CO ratio from biomass waste. Biomass torrefaction as a pretreatment stage has the ability to help meet global energy demand. Pretreatment of biomass is important for breaking down the rigid structure of lignocellulosic biomass cell walls during transformation. The implementation of novel technology to enhance the method of producing green energy is needed. However, gasification integrated with FT reactor is discussed in this paper. This chapter highlights the surplus biomass available, kinetics in biomass gasification, advances in the pretreatment techniques involved in the gasification process, and also catalyst involved in FT reactor.

References

[1] Sieminsky A. International energy outlook; 2014.
[2] Nutongkaew P, Waewsak J, Kiratiwibool W, Gagnon Y. Demand and supply of crude palm oil for biodiesel production towards food and energy security. Appl Mech Mater 2016. Available from: https://doi.org/10.4028/www.scientific.net/amm.839.151.

[3] Hu J, Yu F, Lu Y. Application of Fischer-Tropsch synthesis in biomass to liquid conversion. Catalysts 2012. Available from: https://doi.org/10.3390/catal2020303.
[4] Kim K, Kim Y, Yang C, Moon J, Kim B, Lee J, et al. Long-term operation of biomass-to-liquid systems coupled to gasification and Fischer-Tropsch processes for biofuel production. Bioresour Technol 2013. Available from: https://doi.org/10.1016/j.biortech.2012.09.126.
[5] Leibbrandt NH, Aboyade AO, Knoetze JH, Görgens JF. Process efficiency of biofuel production via gasification and Fischer-Tropsch synthesis. Fuel 2013. Available from: https://doi.org/10.1016/j.fuel.2013.03.013.
[6] Tijmensen MJA, Faaij APC, Hamelinck CN, Van Hardeveld MRM. Exploration of the possibilities for production of Fischer Tropsch liquids and power via biomass gasification. Biomass Bioenergy 2002. Available from: https://doi.org/10.1016/S0961-9534(02)00037-5.
[7] Im-orb, Karittha, and Amornchai Arpornwichanop, Integration of Clean and Sustainable Energy Resources and Storage in Multi-Generation Systems, 283–315, 2020. Available from: https://doi.org/10.1007/978-3-030-42420-6_14.
[8] Hunpinyo P, Cheali P, Narataruksa P, Tungkamani S, Chollacoop N. Alternative route of process modification for biofuel production by embedding the Fischer-Tropsch plant in existing stand-alone power plant (10 MW) based on biomass gasification – Part I: a conceptual modeling and simulation approach (a case study in Thai). Energy Convers Manag 2014. Available from: https://doi.org/10.1016/j.enconman.2014.08.016.
[9] Ng KS, Sadhukhan J. Techno-economic performance analysis of bio-oil based Fischer-Tropsch and CHP synthesis platform. Biomass Bioenergy 2011. Available from: https://doi.org/10.1016/j.biombioe.2011.04.037.
[10] Irfan MF, Usman MR, Kusakabe K. Coal gasification in CO_2 atmosphere and its kinetics since 1948: a brief review. Energy 2011. Available from: https://doi.org/10.1016/j.energy.2010.10.034.
[11] Hanaoka T, Hiasa S, Edashige Y. Syngas production by CO_2/O_2 gasification of aquatic biomass. Fuel Process Technol 2013. Available from: https://doi.org/10.1016/j.fuproc.2013.03.049.
[12] Sands RD, Malcolm SA, Suttles SA, Marshall E. Dedicated energy crops and competition for agricultural land. In: Dedicated energy crops and competition for agricultural land; 2017. Economic Research Report Number 223:252445. Available from: https://doi.org/10.22004/ag.econ.
[13] Bellezoni RA, Sharma D, Villela AA, Pereira Junior AO. Water-energy-food nexus of sugarcane ethanol production in the state of Goiás, Brazil: an analysis with regional input-output matrix. Biomass Bioenergy 2018. Available from: https://doi.org/10.1016/j.biombioe.2018.04.017.
[14] Bressanin JM, Klein BC, Chagas MF, Watanabe MDB, de Mesquita Sampaio IL, Bonomi A, et al. Techno-economic and environmental assessment of biomass gasification and Fischer-Tropsch synthesis integrated to sugarcane biorefineries. Energies 2020. Available from: https://doi.org/10.3390/en13174576.
[15] Hiloidhari M, Das D, Baruah DC. Bioenergy potential from crop residue biomass in India. Renew Sustain Energy Rev 2014. Available from: https://doi.org/10.1016/j.rser.2014.01.025.
[16] The biomass assessment handbook: bioenergy for a sustainable environment. Choice Rev Online. <https://doi.org/10.5860/choice.44-5663>; 2007.
[17] Hall DO, Moss PA. Biomass for energy in developing countries. GeoJournal 1983. Available from: https://doi.org/10.1007/BF00191854.
[18] Bergman PCA, Boersma AR, Zwart RWR, Kiel JHA. Torrefaction for biomass co-firing in existing coal-fired power stations. ECN-C-05-013, Energy Research Centre of The Netherlands ECN; 2005.
[19] Sipra AT, Gao N, Sarwar H. Municipal solid waste (MSW) pyrolysis for bio-fuel production: a review of effects of MSW components and catalysts. Fuel Process Technol 2018. Available from: https://doi.org/10.1016/j.fuproc.2018.02.012.
[20] Gupta VG, Tuohy MG, Kubicek CP, Saddler J, Xu F. Bioenergy research: advances and applications. <https://doi.org/10.1016/C2012-0-00025-7>; 2014.
[21] Molino A, Chianese S, Musmarra D. Biomass gasification technology: the state of the art overview. J Energy Chem 2016. Available from: https://doi.org/10.1016/j.jechem.2015.11.005.

[22] Kumar A, Jones DD, Hanna MA. Thermochemical biomass gasification: a review of the current status of the technology. Energies 2009;2:556–81. Available from: https://doi.org/10.3390/en20300556.
[23] Sindhu R, Binod P, Pandey A. Biological pre-treatment of lignocellulosic biomass—an overview. Bioresour Technol 2016. Available from: https://doi.org/10.1016/j.biortech.2015.08.030.
[24] Chew JJ, Doshi V. Recent advances in biomass pre-treatment—torrefaction fundamentals and technology. Renew Sustain Energy Rev 2011. Available from: https://doi.org/10.1016/j.rser.2011.09.017.
[25] Tumuluru JS, Sokhansanj S, Hess JR, Wright CT, Boardman RD. A review on biomass torrefaction process and product properties for energy applications. Ind Biotechnol 2011. Available from: https://doi.org/10.1089/ind.2011.7.384.
[26] Mei Y, Che Q, Yang Q, Draper C, Yang H, Zhang S, et al. Torrefaction of different parts from a corn stalk and its effect on the characterization of products. Ind Crop Prod 2016. Available from: https://doi.org/10.1016/j.indcrop.2016.07.021.
[27] Dong K, Hochman G, Zhang Y, Sun R, Li H, Liao H. CO_2 emissions, economic and population growth, and renewable energy: Empirical evidence across regions. Energy Econ 2018. Available from: https://doi.org/10.1016/j.eneco.2018.08.017.
[28] Mitsui K, Inagaki T, Tsuchikawa S. Monitoring of hydroxyl groups in wood during heat treatment using NIR spectroscopy. Biomacromolecules 2008. Available from: https://doi.org/10.1021/bm7008069.
[29] Prado KS, Jacinto AA, Spinacé MAS. Cellulose nanostructures extracted from pineapple fibres. Green Energy Technol 2020. Available from: https://doi.org/10.1007/978-981-15-1416-6_10.
[30] Kopczyński M, Plis A, Zuwała J. Thermogravimetric and kinetic analysis of raw and torrefied biomass combustion. Chem Process Eng 2015. Available from: https://doi.org/10.1515/cpe-2015-0014.
[31] Kasim NN, Mohamed AR, Ishak MAM, Ahmad R, Nawawi WI, Ali SN, et al. The effect of demineralization and torrefaction consequential pre-treatment on energy characteristic of palm empty fruit bunches. J Therm Anal Calorim 2019. Available from: https://doi.org/10.1007/s10973-019-08206-8.
[32] Melkior T, Jacob S, Gerbaud G, Hediger S, Le Pape L, Bonnefois L, et al. NMR analysis of the transformation of wood constituents by torrefaction. Fuel 2012. Available from: https://doi.org/10.1016/j.fuel.2011.06.042.
[33] Lam SS, Tsang YF, Yek PNY, Liew RK, Osman MS, Peng W, et al. Co-processing of oil palm waste and waste oil via microwave co-torrefaction: a waste reduction approach for producing solid fuel product with improved properties. Process Saf Environ Prot 2019. Available from: https://doi.org/10.1016/j.psep.2019.05.034.
[34] Chen WH, Peng J, Bi XT. A state-of-the-art review of biomass torrefaction, densification and applications. Renew Sustain Energy Rev 2015. Available from: https://doi.org/10.1016/j.rser.2014.12.039.
[35] Chen Q, Zhou JS, Liu BJ, Mei QF, Luo ZY. Influence of torrefaction pre-treatment on biomass gasification technology. Chin Sci Bull 2011. Available from: https://doi.org/10.1007/s11434-010-4292-z.
[36] Chen WH, Lin BJ, Colin B, Chang JS, Pétrissans A, Bi X, et al. Hygroscopic transformation of woody biomass torrefaction for carbon storage. Appl Energy 2018. Available from: https://doi.org/10.1016/j.apenergy.2018.09.135.
[37] Andersson M, Tillman A-M. Acetylation of jute: effects on strength, rot resistance, and hydrophobicity. J Appl Polym Sci 1989. Available from: https://doi.org/10.1002/app.1989.070371214.
[38] Jędrzejczyk M, Soszka E, Czapnik M, Ruppert AM, Grams J. Physical and chemical pre-treatment of lignocellulosic biomass. Second Third Gener Feed Evol Biofuels 2019. Available from: https://doi.org/10.1016/B978-0-12-815162-4.00006-9.
[39] Schell DJ, Harwood C. Milling of lignocellulosic biomass. Appl Biochem Biotechnol 1994. Available from: https://doi.org/10.1007/bf02941795.

[40] Sankaran R, Parra Cruz RA, Pakalapati H, Show PL, Ling TC, Chen WH, et al. Recent advances in the pre-treatment of microalgal and lignocellulosic biomass: a comprehensive review. Bioresour Technol 2020. Available from: https://doi.org/10.1016/j.biortech.2019.122476.

[41] Zabed HM, Akter S, Yun J, Zhang G, Awad FN, Qi X, et al. Recent advances in biological pre-treatment of microalgae and lignocellulosic biomass for biofuel production. Renew Sustain Energy Rev 2019. Available from: https://doi.org/10.1016/j.rser.2019.01.048.

[42] Asgher M, Ahmad Z, Iqbal HMN. Alkali and enzymatic delignification of sugarcane bagasse to expose cellulose polymers for saccharification and bio-ethanol production. Ind Crop Prod 2013. Available from: https://doi.org/10.1016/j.indcrop.2012.10.005.

[43] Binod P, Janu KU, Sindhu R, Pandey A. Hydrolysis of lignocellulosic biomass for bioethanol production. Biofuels 2011. Available from: https://doi.org/10.1016/B978-0-12-385099-7.00010-3.

[44] Vanegas CH, Hernon A, Bartlett J. Enzymatic and organic acid pre-treatment of seaweed: effect on reducing sugars production and on biogas inhibition. Int J Ambient Energy 2015. Available from: https://doi.org/10.1080/01430750.2013.820143.

[45] Kim KH, Choi IS, Kim HM, Wi SG, Bae HJ. Bioethanol production from the nutrient stress-induced microalga Chlorella vulgaris by enzymatic hydrolysis and immobilized yeast fermentation. Bioresour Technol 2014. Available from: https://doi.org/10.1016/j.biortech.2013.11.059.

[46] Carrillo-Reyes J, Barragán-Trinidad M, Buitrón G. Biological pre-treatments of microalgal biomass for gaseous biofuel production and the potential use of rumen microorganisms: a review. Algal Res 2016. Available from: https://doi.org/10.1016/j.algal.2016.07.004.

[47] Sánchez C. Lignocellulosic residues: biodegradation and bioconversion by fungi. Biotechnol Adv 2009. Available from: https://doi.org/10.1016/j.biotechadv.2008.11.001.

[48] Wan C, Li Y. Fungal pre-treatment of lignocellulosic biomass. Biotechnol Adv 2012. Available from: https://doi.org/10.1016/j.biotechadv.2012.03.003.

[49] Rouches E, Herpoël-Gimbert I, Steyer JP, Carrere H. Improvement of anaerobic degradation by white-rot fungi pre-treatment of lignocellulosic biomass: a review. Renew Sustain Energy Rev 2016. Available from: https://doi.org/10.1016/j.rser.2015.12.317.

[50] Sansaniwal SK, Rosen MA, Tyagi SK. Global challenges in the sustainable development of biomass gasification: an overview. Renew Sustain Energy Rev 2017. Available from: https://doi.org/10.1016/j.rser.2017.05.215.

[51] Gaurav N, Sivasankari S, Kiran GS, Ninawe A, Selvin J. Utilization of bioresources for sustainable biofuels: a review. Renew Sustain Energy Rev 2017. Available from: https://doi.org/10.1016/j.rser.2017.01.070.

[52] Beenackers A, Van Swaaij WPM. Gasification of biomass, a state of the art review; 1984.

[53] Gautam G, Adhikari S, Thangalazhy-Gopakumar S, Brodbeck C, Bhavnani S, Taylor S. Tar analysis in syngas derived from pelletized biomass in a commercial stratified downdraft gasifier. BioResources 2011. Available from: https://doi.org/10.15376/biores.6.4.4653-4661.

[54] Samiran NA, Mohd Jaafar MN, Chong CT, Jo-Han N. A review of palm oil biomass as a feedstock for syngas fuel technology. J Teknol 2015. Available from: https://doi.org/10.11113/jt.v72.3932.

[55] Puig-Arnavat M, Bruno JC, Coronas A. Review and analysis of biomass gasification models. Renew Sustain Energy Rev 2010. Available from: https://doi.org/10.1016/j.rser.2010.07.030.

[56] Golden T, Reed B, Das A. Handbook of biomass downdraft gasifier engine systems. SERI United States Department of Energy, <https://doi.org/10.2172/5206099>; 1988.

[57] Bhavanam A, Sastry RC. Biomass gasification processes in downd raft fixed bed reactors: a review. Int J Chem Eng Appl 2011. Available from: https://doi.org/10.7763/ijcea.2011.v2.146.

[58] Palmer KD, Severy MA, Chamberlin CE, Eggink AJ, Jacobson AE. Performance analysis of a biomass gasifier genset at varying operating conditions. Appl Eng Agric 2018. Available from: https://doi.org/10.13031/aea.12414.

[59] Chawdhury MA, Mahkamov K. Development of a small downdraft biomass gasifier for developing countries. J Sci Res 2010. Available from: https://doi.org/10.3329/jsr.v3i1.5613.

[60] Gil J, Corella J, Aznar MP, Caballero MA. Biomass gasification in atmospheric and bubbling fluidized bed: effect of the type of gasifying agent on the product distribution. Biomass Bioenergy 1999. Available from: https://doi.org/10.1016/S0961-9534(99)00055-0.

[61] Baskoro AN, Aptari OE. Biomass waste and low rank coal gasification technology with carbon capture system to optimize a clean energy production as an alternative solution to achieve energy security in Indonesia. Indones J Energy 2020. Available from: https://doi.org/10.33116/ije.v3i2.90.

[62] Situmorang YA, Zhao Z, Yoshida A, Abudula A, Guan G. Small-scale biomass gasification systems for power generation (<200 kW class): a review. Renew Sustain Energy Rev 2020. Available from: https://doi.org/10.1016/j.rser.2019.109486.

[63] Dassey A, Mukherjee B, Sheffield R, Theegala C. Catalytic cracking of tars from biomass gasification. Biomass Convers Biorefinery 2013. Available from: https://doi.org/10.1007/s13399-012-0063-1.

[64] Martínez JD, Mahkamov K, Andrade RV, Silva Lora EE. Syngas production in downdraft biomass gasifiers and its application using internal combustion engines. Renew Energy 2012. Available from: https://doi.org/10.1016/j.renene.2011.07.035.

[65] Im-orb K, Simasatitkul L, Arpornwichanop A. Techno-economic analysis of the biomass gasification and Fischer-Tropsch integrated process with off-gas recirculation. Energy 2016. Available from: https://doi.org/10.1016/j.energy.2015.11.012.

[66] Pradhan P, Arora A, Mahajani SM. A semi-empirical approach towards predicting producer gas composition in biomass gasification. Bioresour Technol 2019. Available from: https://doi.org/10.1016/j.biortech.2018.10.073.

[67] Surjosatyo A. Tar content evaluation of produced gas in downdraft biomass gasifier. Iran J Energy Environ 2012. Available from: https://doi.org/10.5829/idosi.ijee.2012.03.03.1588.

[68] Martinelli M, Gnanamani MK, LeViness S, Jacobs G, Shafer WD. An overview of Fischer-Tropsch synthesis: XtL processes, catalysts and reactors. Appl Catal A Gen 2020. Available from: https://doi.org/10.1016/j.apcata.2020.117740.

[69] Botes FG, Niemantsverdriet JW, Van De Loosdrecht J. A comparison of cobalt and iron based slurry phase Fischer-Tropsch synthesis. Catal Today 2013. Available from: https://doi.org/10.1016/j.cattod.2013.01.013.

[70] O'Brien RJ, Xu L, Bao S, Raje A, Davis BH. Activity, selectivity and attrition characteristics of supported iron Fischer-Tropsch catalysts. Appl Catal A Gen 2000. Available from: https://doi.org/10.1016/S0926-860X(99)00462-7.

[71] Chonco ZH, Lodya L, Claeys M, Van Steen E. Copper ferrites: a model for investigating the role of copper in the dynamic iron-based Fischer-Tropsch catalyst. J Catal 2013. Available from: https://doi.org/10.1016/j.jcat.2013.08.012.

[72] Cook KM, Perez HD, Bartholomew CH, Hecker WC. Effect of promoter deposition order on platinum-, ruthenium-, or rhenium-promoted cobalt Fischer-Tropsch catalysts. Appl Catal A Gen 2014. Available from: https://doi.org/10.1016/j.apcata.2014.05.013.

[73] Van Der Laan GP, Beenackers AACM. Kinetics and selectivity of the Fischer-Tropsch synthesis: a literature review. Catal Rev Sci Eng 1999. Available from: https://doi.org/10.1081/CR-100101170.

[74] Rochana P, Wilcox J. A theoretical study of CO adsorption on FeCo(100) and the effect of alloying. Surf Sci 2011. Available from: https://doi.org/10.1016/j.susc.2011.01.003.

[75] Yang J, Qi Y, Zhu J, Zhu YA, Chen D, Holmen A. Reaction mechanism of CO activation and methane formation on Co Fischer-Tropsch catalyst: a combined DFT, transient, and steady-state kinetic modeling. J Catal 2013. Available from: https://doi.org/10.1016/j.jcat.2013.05.018.

[76] Ma W, Shafer WD, Martinelli M, Sparks DE, Davis BH. Fischer-Tropsch synthesis: using deuterium tracer coupled with kinetic approach to study the kinetic isotopic effects of iron, cobalt and ruthenium catalysts. Catal Today 2020. Available from: https://doi.org/10.1016/j.cattod.2019.01.059.

[77] Ojeda M, Nabar R, Nilekar AU, Ishikawa A, Mavrikakis M, Iglesia E. CO activation pathways and the mechanism of Fischer-Tropsch synthesis. J Catal 2010. Available from: https://doi.org/10.1016/j.jcat.2010.04.012.

[78] Ertl G, Knözinger H, Weitkamp J. Handbook of heterogeneous catalysis. <https://doi.org/10.1524/zpch.1999.208.part_1_2.274>; 2008.

[79] Önsan ZI, Avci AK. Multiphase catalytic reactors: theory, design, manufacturing, and applications. <https://doi.org/10.1002/9781119248491>; 2016.

[80] Steynberg AP, Dry ME, Davis BH, Breman BB. Fischer-Tropsch reactors. Stud Surf Sci Catal 2004. Available from: https://doi.org/10.1016/s0167-2991(04)80459-2.

[81] Bukur DB, Ma WP, Carreto-Vazquez V. Attrition studies with precipitated iron Fischer-Tropsch catalysts under reaction conditions. Top Catal 2005. Available from: https://doi.org/10.1007/s11244-005-2885-6.
[82] Van Steen E, Schulz H. Polymerisation kinetics of the Fischer-Tropsch CO hydrogenation using iron and cobalt based catalysts. Appl Catal A Gen 1999. Available from: https://doi.org/10.1016/S0926-860X(99)00151-9.
[83] Deckwer WD, Serpemen Y, Ralek M, Schmidt B. Modeling the Fischer-Tropsch synthesis in the slurry phase. Ind Eng Chem Process Des Dev 1982. Available from: https://doi.org/10.1021/i200017a006.
[84] Sansaniwal SK, Pal K, Rosen MA, Tyagi SK. Recent advances in the development of biomass gasification technology: a comprehensive review. Renew Sustain Energy Rev 2017;72:363–84. Available from: https://doi.org/10.1016/j.rser.2017.01.038.
[85] Parajuli PB, Deng Y, Kim H, Yu F. Cost analysis model for syngas production cost evaluation using the graphical user interface. Energy Power 2014;4.
[86] Dell'Antonia D, Cividino SRS, Malev O, Pergher G, Gubiani R. A techno-economic feasibility assessment on small-scale forest biomass gasification at a regional level. Appl Math Sci 2014. Available from: https://doi.org/10.12988/ams.2014.46450.
[87] Andersson J, Lundgren J. Techno-economic analysis of ammonia production via integrated biomass gasification. Appl Energy 2014. Available from: https://doi.org/10.1016/j.apenergy.2014.02.029.
[88] Muresan M, Cormos CC, Agachi PS. Techno-economical assessment of coal and biomass gasification-based hydrogen production supply chain system. Chem Eng Res Des 2013. Available from: https://doi.org/10.1016/j.cherd.2013.02.018.
[89] Elkadeem MR, et al. Feasibility analysis and techno-economic design of grid-isolated hybrid renewable energy system for electrification of agriculture and irrigation area: A case study in Dongola, Sudan. Energy Conversion and Management 2019;196:1453–78. Available from: https://doi.org/10.1016/j.enconman.2019.06.085 In this issue.

CHAPTER 4

Energy recovery from biomass through gasification technology

4.1 Introduction

The process of the thermochemical conversion (TCC) process for converting the carbonaceous matters to useful by-products is termed gasification. It has an important role in the formation of the by-products like fuels or as a chemical feedstock and also the best method for the generation of syngas, which is a combination of H_2 and CO along with little quantity of CO_2, methane gas, and nitrogen content [1]. The gasification method was mainly used in the formation of the oil with the help of Fischer—Tropsch (FT) synthesis along with coal gas as feedstock that was then used at a high level to drop the expectations on oil resources that are imported. Ecological issues like global warming rise in climatic concerns to create a neat atmosphere have led to the production of the integrated gasification combined cycle (IGCC) which is used for generating power for gaining neat fuels. This causes the refiners to have decided to make utilization of the gasification plants for producing useful products. Biomass is utilized to get energy sources mainly that have heat and electricity [1]. India is well-known for having extensive coal deposits across the world, mainly United States with 250,016 MT and a percentage of 24.2%, which is followed by Russia with 160,364 MT and a percentage of 15.5%, then Australia with 144,818 MT and a percentage of 14%, and China with 138,819 MT and a percentage of 13.4% till the month of December in the year 2017, having nearly 97,728 MT of deposits were found with percentage of 9.4% and hence can be used on a larger scale for the generation of power with advanced gasification method [2]. Recently, the gasification methods used are very important in gaining products, such as biofuels, biomethane, and H_2, which are considered as potential renewable fuels. There is a rise in the generation of power using gas turbines that runs on syngas generated using the process of gasification of biomass. Hence, the way to give a process with high energy efficiency and sustainability is by focusing on syngas generation developing the quality of gas and the purity of gas and improving the efficiency of the process and its expense for production. The gasifier used is highly influenced by the properties and methods of preparation of the biomass. Each variety of feedstock requires a specified conversion step of energy with some known design parameters. Gasifier reactors are varied based on many factors like the supply of heat as in all thermal gasification and autothermal gasification and properties as in entrained flow, fluidized bed, fixed bed, agent for gasification process,

A Thermo-Economic Approach to Energy From Waste
DOI: https://doi.org/10.1016/B978-0-12-824357-2.00007-3

such as air, steam, and O_2, and pressure in the gasifier, which are atmospheric pressure and, vacuum pressure for the transportation [3]. It has been investigated that many gasification setup have packed bed gasifier and also the fluidized bed gasifier (FLBG) and the entrained suspension coal gasifier [3]. The importance of renewable energy/fuel in today's world is demonstrated here, along with the foundation to a generation of biomass and properties of plants linked with them is explained. Various methods are used to convert energy, and gasification methods are mainly used here as it has many uses and is highly important. Also, many other gasifiers are also studied on an important basis [4]. The importance of renewable energy/fuel in today's scenario is highlighted in this chapter, which is followed by a brief overview of biomass gasification and the principle of anaerobic digestion with parameters affecting the process. Subsequently, different types of gasifiers are critically analyzed. Various energy conversion methods are discussed, among which gasification technology has been predominantly chosen for a detailed discussion in this chapter with life cycle and aspen plus approach.

4.2 Thermochemical conversion

This conversion method is mainly utilized to convert heat with or without oxygen to get energy from the biomass.

4.2.1 Combustion

The process of heating the feedstock with the air for producing energy and power obtained in the form of chemical energy through gasification of the biomass is termed the process of combustion. At a heated temperature between 800°C and 1000°C, hot gases begin to form from the biomass [1]. Most of the feedstocks are burnable, but it may not be possible to put them in action if the water content of the biomass is greater than 50%. Having more water content in biomass is more comfortable with biochemical conversion (BCC) when compared with TCC [5]. When coal and natural gas and other fossil fuels are burnt with the feedstock for forming the energy is described to be the method of cofiring. There are lots of benefits, mainly when power is the main output. Nowadays the fossil fuel power plant is modified so that it is utilized for cofiring the biomass for energy production. There is a large drop in the release of sulfur, CO_2, and other harmful products as the biomass is used in a modified gasifier plant [5].

4.2.2 Pyrolysis

The term "pyrolysis" is a mixture of two Greek words, "pyro" and "lysis," which says pyro is meant for "fire," and lysis is meant for "separation." It is a method in which the feedstock is influenced by high temperature (around 500°C) with or without O_2 in pressurized conditions. Biochar, which is residues in solid and other fuels in a liquid state, is formed due to incomplete burning of the biomass. Liquid fuels are described as biofuel or bio-crude, which are formed mainly when it has an efficiency of nearly 80%, that are influenced to flash pyrolysis with low heat input [6]. At the same time, biochar is mainly charcoal that is high in the carbon content and is utilized for improving the soil properties and many other helpful uses. These outputs are high in density in energy than the primary biomass feedstock that helps in a high drop of the economy for transportation [6].

4.2.3 Gasification

Gasification is the partial combustion of biomass. It provides flexibility in making use of various types of biomass materials and also in the production of various outputs. Hence, all biomass type is possibly changed into syngas with the help of the gasification process. The syngas consists of H_2, CO, CO_2, and CH_4, which are the bases for all means of energy or carriers of the energy like those of heat, H_2, biomethane, biofuels, power, and also chemicals that are given [7]. Extraction of FT diesel, methane, methanol, and dimethyl ether is given with processes. The utilization of the given resources of biomass requires a high efficiency and sustainable process. For the utilization of biomass, the gasification provides a high potential along with high efficiency to the process. The process of gasification of biomass is done with the help of the partial oxidation method of carbon, which is in the biomass and is set to a high temperature along with a limited quantity of an oxidant that is air, pure O_2 or H_2O, or steam. The gasification process id mainly influenced by the temperature in the gasifier, the oxidant used in a gasifier, and the time of residence in the gasifier [7]. The mixtures and characteristics of the syngas are mainly based on the biomass feedstock, the type of gasifier used, and the conditions of the gasifier operations. Gasification combined along with air results in the production of syngas that has a heating value of ranging 4–7 MJ/m^3; at the same time when pure O_2 or H_2O steam is utilized as an oxidant it results in a significant rise in the heating values of the gas nearly in the range of 10–18 MJ/m^3 [7].

Biomass consists of a wide variety of types of biomaterials, such as wood, forest residues, waste from tree wood, agricultural residues, grasses, wastes from the food industry, energy algae, bagasse, sewage sludge, and straw. This utilization of varied biomass leads to many problems and their corrections for introducing the biomass, storing the biomass, transportation, gasifier running, and cleaning the generated syngas. The mostly utilized biomass gasifiers are basically classified as entrained flow gasifiers, moving bed gasifiers, FBGs, and FLBGs [8]. FLBGs and entrained flow gasifiers give an intensive relation for the gas along with the biomass that leads to a rise in the rate of reaction and efficiency of conversion. FBGs mostly consist of transferring lower heat and mass and form a large quantity of tar and char. However, operating conditions and the design are very simplified and not even very big in size [8].

4.2.4 Principles of anaerobic digestion

Anaerobic digestion is one of the types of BCC techniques. Anaerobic digestion is used in biological methanogenesis, where an anaerobic process influences the decomposition of carbonaceous matter in eco-friendly surroundings leading to the accumulation of organic compounds and a drop in O_2 for aerobic metabolism. This phase is done with the help of many other microbes that are available plenty in nature, such as intestines of the animals, soils that are collected from floods, sediments, and landfills. Anthropogenic caused due to humans leads to the generation of methane; leaving it into the air is the main problem as it releases CO_2, which is the main greenhouse gas (GHG). The main problem is due to crops in the flooded soil, facilities for handling animal wastes, landfills, and domestic animals that have rumens [9].

In anaerobic digestion, the biomass is collected, followed by heavily shredding and keeping in a reactor that has more microbes needed in methane fermentation. A conventional reactor is combined, supplied biomass once or more per day, heated to 35°C, and run at a hydraulic retention time of 20–30 days and a rate of loading up to 1.7 kg volatile solid (VS) (organic matter is considered as ash-free dry weight) m^3/day (0.1 lb VS ft day). There is nearly a 60% drop in organic matter gained with respect to the yield of methane of 0.24 m^3 per kg (4.0 ft per lb). The composition of biomass has 60% of CH_4 and 40% of CO_2 with some hydrogen sulfide and H_2O vapor. Solid residues get deposited or dewatered in other ways and utilized as compost. The syngas gained can be utilized for eradicating CO_2 and hydrogen sulfide [9].

This traditional design is changed by many new designs that are depending mainly on the feed suspended solid content. The main aim of these new designs is to raise the solids and microbe's retention, drop in the size of the reactor, and drop the energy needs of the process. For dilute low solids (<1%), waste products, such as food processing wastes and attached-film reactors, are kept in action. Interaction of organisms onto inert media allows less time of retention, which is <1 day excluding washout. In designs for feeds with medium solids (5%–10%), matters, such as sewage sludges or aquatic plants, solids, and organic matters, are recycled looking after the deposition of matters in the digester or in other digesters [10]. For high-solid biomass that has >10%, the methods, such as high-solid stirred digesters or leach bed batch systems, are implemented. These innovative designs have increased loading rate by 20 times, reduced residence time, and improved in-process stability. Biochemical CH_4 research reveals that biomass feedstocks have a number of advantages (Table 4.1) may rise by more than 50%

Table 4.1 Chemical composition of biomass.

Biomass name	Lignin (%)	Hemicellulose (%)	Cellulose (%)
Soft wood	30	30	40
Hardwood	25	25	50
Switchgrass	20	35	45
Wheat straw	25	25	45

generally linked to the anaerobic digestion; like the biodegradability of any sorghum species that has more than 90% with respect to the yield of CH_4 of 0.39 m^3 per kg VS [10].

4.3 Production and use of aquatic biomass

Many wastes obtained from the biomass are generated in all over the year across the globe [11]. Recently, it was burned or transformed into organic fertilizers in natural means in some given conditions. But biomass waste has turned to be a rising issue nowadays, and it also leads to many ecological issues [12]. If we consider the combustion process of agricultural waste seen in many backward nations doing even now, it causes air pollution [13]. This leads to eradication of the waste outputs while burning of wastes of biomass into the air, such as CO, NO_x, NO_2, and other matters. These waste-polluting substances are supported by the generation of O_3 gas and HNO_3 acid, leading to the settling of acid, which leads to danger to man life and ecological health [13]. Also, wastes from fruit and vegetable are mainly formed during the generation and storage phases due to over formation due to variations in the climate, poor skills, disasters due to nature, and a poor structure [14]. A high amount of wastes are thrown into landfills, causing the production of CH_4 and CO_2, the contamination of water on the surface, the contamination of groundwater, smell, and the contamination of land. CH_4 released at landfills is a crucial donator to GHGs, due to which the potential of global warming rises. Nearly 60% of the emissions of CH_4 all over the world are from farming, wastewater, the generation and transport of fossil fuels, and landfills, as stated by Singh et al. [15]. Moreover, leachate obtained in landfill has a large amount of NH_3-N, some dangerous matters, heavy metals, and organics. It affects the plant growth. The utilization of leachate in irrigation modifies the properties of soil, such as salinity and biotoxicity [13]. Thus the waste of feedstock can lead to big problems, such as pollution and ecological problems linked to accumulation, decomposition of biomass, and treatment of biomass.

4.3.1 Potential of biomass waste

The total biomass waste is formed from farming, and forest, processing of food, and other parts of the industry are rising due to more people and rising industries. Biomass waste is commonly found in farming since it eliminates wastes like straws of rice, corn, and also of sugarcane leftover bagasse, wheat, and rice husk, leading to produce 731, 354, 204, 181, and 110 Mt, respectively, every year stated by Sarkar et al. [16]. These quantities of wastes from biomass are generally thrown. Residues from forestry are formed by harvesting, and the processing of product is shown nearly as 72.5 Mt in the United States and Canada as studied by Koutinas et al. [17]. The food sector has an equal responsibility to make large biomass waste, with 35 Mt of rapeseed meal, 5–9 Mt of grape pomace, 15.6 Mt of citrus waste, 3–4.2 Mt of apple pomace, 9 Mt of banana waste, which are formed all over the globe in 1 year as investigated by Djilas et al. [18] and Padam et al. [19]. Waste formed from the industry of olive oil leads to big ecological issues, mainly in the Mediterranean where nearly 30 Mt of waste are formed in 1 year as studied by Caputo et al. [20]. The coffee sector gives nearly 7.4 Mt and an extra huge quantity of cherry husk, silver skin, and coffee pulp, which can destroy the ecosystems as they have the decomposition qualities of organic substances [21]. Hence, more biomass waste can be used for giving a solution to the disposal issues. Waste from biomass is very varied with respect to chemical mixing and has shown significant potential in the improvement of cost-efficient biorefineries. Residues from farming and forest consist of many cellulose and hemicellulose, as stated in Table 4.1, that are mainly used in forming fermentable sugars and biofuels, which are ethanol and butanol. Waste from biomass obtained from sectors like processing of food that processes carrots and apple/pear pomaces has cellulose and hemicellulose contents in huge quantities and hence can be used for sugar conversion without the necessity of difficult pretreatment steps stated by Nawirska and Kwaśniewska [22]. The coffee grounds have a large amount of hemicellulose that has 30.1% mannan as the main part in residue with polysaccharide 19.3%; this shows that spent grounds are very good products for the generation of useful mannose and manno-oligosaccharides investigated by Nguyen et al. [23]. Waste from the biomass sources that are high in xylan content is found in the corn cobs, switchgrass, wheat straw, rice straw, sugarcane bagasse, and corn stover that have the capacity to be a raw material sources for xylooligosaccharides generation. Pectic oligosaccharides that have raised as a new class of prebiotics are gained from the residues of pectin that has agriculture residues, such as olive pomace (34.4%), onion skin (27%–34%), apple pulp (20.9%), citrus waste (30%), soy hull (16.3%), sugar beet pulp (16.2%), and potato pulp (15%), stated by Babbar et al. [24]. Flavonoids, phenolic acids, carotenoids, and their derivatives are the main bioactive matters, which are seen in wastes of fruit and vegetable, showing antiinflammatory, antioxidant, antiatherogenic, antimicrobial, antiallergenic, and antithrombotic works with active utilization.

4.4 Lignocellulose biomass pretreatment

The following sections show the new developments in various processes of pretreatment shown in the literature. The investigation mainly looks at showing various ways to utilize the pretreatment process for different ideal strategies that are included in the process of converting ethanol from the LCM. These pretreatment processes are also the main primary phases in the lignocellulosic biorefinery [25].

4.4.1 Physical methods

The heating of the substrate and pressuring it can be achieved with the help of high energy. The thermal pretreatment makes use of heat that breaks down the bonds in matters that raise the process of digestion. Heating is also achieved with the help of steam explosion too. But there are times when some biodegradable matters can reduce the volatile organics, leading to a drop in biogas [26]. Thermal pretreatment consists of microwave thermal treatment, steam explosion treatment, and thermal hydrolysis treatment. During the pretreatment by thermal method, the biogas generation is raised with the help of partial solubilization of compounds of the inorganic and organic present. Bien et al. investigated that the rise in the generation of CH_4 to nearly 25% also rose in the sewage sludge by thermal pretreated in the temperature range of 170°C–180°C with respect to the untreated sludge. Chemical pretreatment methods have phases of oxidation, acid pretreatment, ozonolysis, and alkaline uses. This leads to rising the solubility and natural decomposition of chemical mixtures of biomass. According to Moset et al. [27] acid treatment pig slurry with H_2SO_4 yields around 20% more CH_4 than untreated slurry. Mechanical pretreatment also focuses on the drop of the size of a particle by giving force in many ways in developing the process of digestion by the step of transferring the heat and transferring of the mass [26].

In the generation of biogas, the main factor that acts as a crucial role is the size of the particle. Microbial working is raised by giving the tiny size of particle that causes a rise in the usage of a substrate. The majority of biogas was produced at particle sizes of 0.40 and 0.088 mm size that caused by the influences of the size of particles on the wastes from the hilly trees and farming that has many particle sizes (0.30, 0. 60, 1.10, 0.40, and 0.088 mm).

Ultrasonic wave is utilized for crushing the cell membrane of the substrate and is required nutrient that will raise the usage of the substrate more than generating biogas, but it is not economical [28]. Ultrasonic pretreatment gives more conditions that are favored to degrade naturally by disintegrating and enhancing the particles by the drop in the biogas. Pretreatment normally drops the volatile solids and raises the generation of biogas in an anaerobic reaction. Braguglia et al. [28] tested the waste sludge pretreated with ultrasonic that gives nearly 30% more biogas than the untreated sludge.

4.4.2 Chemical Methods

To raise the generation of biogas, the chemicals that are utilized for breaking down the links change the substrate to soluble ones. The alkalis, such as sodium hydroxide or potassium hydroxide, and acids, such as H_2SO_4, are utilized based on the substrate of the solution and pH [29]. The speed of digestion matters on the substrate. Alkalis raise the surface area of the digestion. However, this results in the formation of some hazardous chemical agents that have a negative impact on the ecosystem, and chemicals are not acceptable [30].

Chemical pretreatment makes utilization of the compounds, such as oxidants, acids, or alkalis, to break the complex organic seen in biomass to a simplified design [31]. Breaking of the organic structure of the biomass by means of cut on the lignin—carbohydrate bonds and cellulose transparent matrix else on the hydrolysis of hemicellulose are considered to be the most desired functions of chemical pretreatments for the biomass, which is lignocellulosic biomass. Strong acid treatment with the help of H_2SO_4 and HCl kept at high levels leads to the blockage of the process of anaerobic digestion, resulting in many waste by-products, such as furfural and its derivatives [32], as there is more degradation of a complex substrate when testing with strong acids leads to the drop of fermentable sugar. Thus for the process of pretreatment with acids, the most used acid is diluted acids. Acid pretreatment is combined with large temperatures, which are greater than 1000°C. Alkali pretreatment is mostly utilized with respect to acid pretreatment as it shows circumstances for anaerobic digestion that are supportive by stopping the fall in pH value [33]. Alkali pretreatment with sodium hydroxide or potassium hydroxide is done in normal air temperature or more nearly 40°C that has a long reaction time [34]. When an alkali agent is NH_4, more temperature and pressure are required for dropping the pretreatment time along with the benefits of ammonia freeze explosion [35].

4.4.3 Biological pretreatment

Brown, white, and soft rot fungi are used in the biological pretreatment to depolymerize lignin and hemicelluloses in LCM [36]. Brown rots majorly break the defense of cellulose. At the same time, white and soft rots strike cellulose and lignin [37]. Fungi turn lignin with the help of an anaerobic reaction that makes use of many extracellular enzymes, namely ligases. Along with the nature of the biomass, the composition and other process factors, such as type of microbes in process, temperature of incubation, pH, time of incubation, inoculums level, water content, and rate of aeration, influence the performance of the process of biological pretreatment [38].

4.5 Bioconversion and downstream processing of biomass-derived molecules' conversion to chemicals

Biomass is the recently found renewable material on the globe. It is seen that nearly the total biomass formed is 130 billion tons annually [39]. With the help of biological and/or thermochemical methods, the biomass can be changed to some chemicals, fuels, and other substances [40]. The International Energy Agency shows that nearly 10% of the global energy is given by biomass by 2035 and nearly 27% of vehicle fuel is altered by energy obtained from the biomass by 2050 [41]. Energy gained from biomass is an eco-benign process that has less action on the ecosystem for the carbon-neutral parameters. Biomass may be converted into a variety of materials, including chemicals, fuels, and carbon, through a variety of processes, including biochemical, thermochemical, and physiochemical [42]. Waste biomass is also utilized to generate the biochar, that is an eco-friendly and sustainable material used in the soil amendment, synthesis of chemical, treatment of wastewater, and recovery of energy [43]. For improvement in the properties of the biochar and its performance, many types of research are done to make it economical for biochar synthesis and its changes like impregnation of metals [44]. With thermochemical methods, it is possible to form many useful outputs from the feedstock. A big summary on the valorization of biomass includes the pretreatment of the feedstock, process of the catalytic reaction of biomass to form chemicals, synthesis of the carbon obtained from biomass, and uses of carbons gained from the biomass.

4.6 Energy recovery for heating or process applications

4.6.1 Steam cycle

The easiest method for energy recovery is the steam cycle. As no pretreatment of gases is processed, the tar is allowed to undergo combustion, and no effect on the boiler is seen by studies done by Quaak et al. [45]. The highest efficiency of the electricity in steam cycle gasifier plant is nearly 23%, which is high with respect to the traditional solid waste incinerator's efficiency [46]. There are some restrictions for the incineration of the waste, and the steam cycle gasifier boiler is the highest temperature of metals for the superheater tubes, as it is set to below 450°C to stop the high rate of corrosion in the tubes caused by hydrochloric acid, which are seen in the flue gas. This leads to restrictions to drop the temperature in steam used in the steam turbine that causes a reduction in the electrical efficiency of the overall gasifier [46]. These restrictions can be treated with the help of gas pretreatment or by combining them with a thermoelectric power plant in a gasification—steam cycle plant.

The pretreatment process for gas can eliminate the hydrochloric acid prior to entering the burner, leading to the burning of the pure gas in the mixed setup of the modern boiler that gives an allowance of the temperature of about 520°C and a rise of 6% in electrical efficiency investigated. The combination along with traditional power plants is known as "cofiring": it permits a rise in the performance that has benefits of the large-scale efficiency of the steam cycle in thermoelectric power plant. Mainly, a cofiring system is done in two different configurations: taking a gas burner in a different boiler for the water vaporization step and taking a burner that burns gas in the boiler as the main fuel [46].

4.6.2 Engine

Spark ignition engines usually consumed along with the help of petrol or kerosene can be worked with gas alone. Diesel engines are used to transform to operate on full gas by dropping the pressurized compression ratio and by using a spark ignition system. Due to lower heating value (LHV), engines producing gas have low efficiency than those not transformed; but modifications in a modern engine can make up to nearly 25% of the net output of electricity. The engines show benefits of being robust and sustainability to the wastes generated than gas turbines investigated by Bridgwater. The same situation may arise in the gas turbine when the turbocharger is used to pressurize the gas in it stated by Bridgwater. The major limitation of gas engines is the low gain in the efficiency combined with combined-cycle mode and the low economy efficient scale by Bridgwater [47].

4.6.3 Gas turbine

The turbines of the advanced combined cycle are the most successful dependable power plant that has a nearly 60% of efficiency rate [48]. There is a drop of up to 40% of the net output of electricity used effectively in the pretreatment of the produced gas. Even, gas turbines show sensitivity to the quality expected in the gas, only very low levels of wastes, mainly tar, alkali metals, S, and Cl, can be exempted, reported by Bridgwater [47]. The chemical recovery cycle is the latest and best option. Here, the energy in the turbine exhaust gas is utilized to provide the process of pretreatment of the gas, such as catalytic cracking of tar or steam reforming reported by Heppenstall [49]. Typical gas turbines have been used in the low LHV: for a simple start-up step, the burners should be used for dual fuel operation, and longer combustion chambers are required for the control of CO emission.

4.6.4 Biogas

The traditional fuel resource availability, rise in fuel economy, and the rising problem for ecological concerns have led scientists to introduce new methods to gain clean and sustainable energy with the help of renewable energy sources. The biogas technique shows an interesting method for using biomass sources for making the needs of partial energy. A biogas system can have many advantages to the user and guards the ecosystem. This system forms biogas, a mixture of CH_4 and CO_2 with a small number of different gases in the organic matters using anaerobic digestion. In anaerobic digestion, it is seen that the organic matter is degraded with the help of microbes without O_2. A complex reaction occurs that converts many polymeric matters, such as proteins, carbohydrates, and lipids, that has C atoms at many oxidation phases and/or reduction phases, to one-carbon matters are mostly in the oxidized state of carbon dioxide (CO_2) or either in the reduced state of methane gas (CH_4) [50].

4.7 Conversion of lignocellulosic biomass–derived intermediates lignin biorefinery biogas from waste biomass

The biochemical process, also called as anaerobic digestion process, undergoes four phases that are methanogenesis, acetogenesis, acidogenesis, and hydrolysis. The second and third phases are stages where acids are formed, and the last phase is called stages where methane is formed [51]. The reaction involved and the microbes used in all four phases are given below.

4.7.1 Hydrolysis

The first phase is called hydrolysis, in the process of anaerobic digestion. Here, the complicated polymers that are organic in nature are transformed into simple soluble substances. At that time, the lipids (fats) are changed to form carbohydrates that are polysaccharides, fatty acids are changed to form simple sugars that are monosaccharides, and proteins are changed to form the amino acids. Hydrolysis is done with the help of many other groups of facultative or obligate fermentative microbes with the support of excreting extracellular enzymes. Lipases are transformed from the lipids to form the long-chain fatty acids, and proteases are transformed from proteins to form the amino acids along with polysaccharides, such as cellulose, starch, and pectin, and undergo hydrolyzation to form monosaccharides with the help of cellulases, amylases, and pectinases [51].

4.7.2 Acidogenesis

Here, the soluble molecules that are formed with the help of hydrolysis are changed to form volatile fatty acids (C_1-C_5), H_2, CO_2, ethanol, and some organic N and S_4. The acids formed here are acetic acid symbolized as CH_3COOH, propionic acid symbolized as CH_3CH_2COOH, butyric acid symbolized as $CH_3CH_2CH_2COOH$, and valeric acid symbolized as $CH_3CH_2CH_2CH_2COOH$. The formed output of acetic acid is taken to the last stage directly, and the other outputs are used in the third stage for further process with the help of acetogens [51].

4.7.3 Acetogenesis

Acetogenesis is termed as the third phase of this process. The volatile fatty acids produced in the preceding phase are converted to H_2, acetic acids, and CO_2 using acetogens [51].

4.7.4 Methanogenesis

This is the last step, in which methanogenic microorganisms, also known as methanogens, convert acetic acid, hydrogen, and CO_2 into CH_4. Nearly 66% of CH_4 is produced from acetic acids with the help of acetate decarboxylation, and balance 34% of CH_4 is produced from the reduction process of CO_2 [51].

4.8 Parameters affecting anaerobic digestion process

Parameters that influence the performances of the anaerobic digestion are given below.

4.8.1 Temperature

Temperature is the main factor that is taken in anaerobic digestion. Many types of methanogens work in three various temperatures, which are 45°C–60°C known as thermophilic, 20°C–45°C known as mesophilic, and below 20°C known as psychrophilic. The rate of generation of biogas rises when there is a rise in temperature. In the process of biogas digestion, only mesophilic temperature and thermophilic temperature are taken as important as the anaerobic digestion reaction normally does not work below 10°C. The microbes that are used for the process of digestion are very delicate to temperature variations so it is required to keep a constant temperature. Thermophilic microbes work efficiently with respect to the time of retention, the rate of loading, and the yield of gas gained but at the same time more amount of heat input and variations in the temperature sensitivity and ecological changes than mesophilic. This has an effect on the temperature affecting the rate of process of anaerobic digestion [52].

4.8.2 Solid to water content

Water and raw material should be mixed at the same time to form a slurry that has needed consistency. Formation of biogas is not efficient if the formation of slurry is very diluted or very thin. The ideal solid concentration is varied in the range between 7% and 25% based on the raw material utilized here. Sewage waste has solid content to be very low, due to which the ideal level can be gained only when substances such as residues of crops plants and weeds are added. Nsair et al. tested and showed the impact of total solid matters with 2.6%, 4.6%, 6.2%, 7.4%, 9.2%, 12.3%, and 18.4%. Total solid on the yield obtained from biogas made use of cattle foods in a batch digester that has a volume of 400 mL and showed that 7.4% and 9.2% of total solids gained more working performance on the yield of biogas when compared to other percentages [52].

4.8.3 pH level

The ideal level of pH in the anaerobic digester is the range of 6.7–7.5. The pH level is not the same in the entire process. The rate of formation of volatile fatty acids is greater than the rate of production of methane, leading to a drop of pH level inhibiting the methanogens, but it is sensitive to the conditions of the acids. Drop in the pH can be managed by adding chemicals, such as ammonium hydroxide, potassium, gaseous NH_4, sodium bicarbonate, lime, Na_2CO_3, and $NaOH_2$. The researcher showed the impact of pH (4, 7, and 9) with the help of cow dung as feed and stated that pH 7 would give a high yield of biogas by 9 and then 4 [52].

4.8.4 Retention period

The retention period is stated as the time where the organic matters stay inside the digester for the production of biogas. The retention period is influenced by the variety of feedstock used and the temperature utilized. The SRT and HRT, also known as the solids retention time and hydraulic retention time, are the main RT in the process of anaerobic digestion. SRT is the time that microbes (solids) stay in the digester. HRT is specified for denoting the SRT. It is described as the time given by the slurry fed in the digester from the point of its entering and to the point of exiting from the gasifier [51].

4.8.5 Organic loading rate

Organic loading rate (OLR) is one of the main factors that influence the production of biogas in anaerobic digestion, especially at the time of digestion in a continuous flow mode. OLR is a measurement for measuring the capacity of biological for anaerobic digestion system. It is shown as the quantity of raw material (kg of volatile solids) used to feed the digester for a unit volume in a day. Excessive loading can be highly influencing the process of digestion because of the collection of acids. The ideal rate of loading is in the range of 0.5 and 2 kg for entire volatile solids in a unit volume of the digester in a day depending on the raw material used, time of retention, and the process temperature [51].

4.8.6 C/N ratio

The carbon and nitrogen seen in the biomass are mentioned by the C/N ratio 15. The carbon–nitrogen (C/N) ratio plays a vital role in the production of biogas. The major consumables for anaerobic bacteria are elements of carbon, which are in the form of carbohydrates, and nitrogen, which are in the form of proteins and ammonia nitrates. Bacteria consume carbon 30 times more likely than nitrogen. To obtain an optimum rate, the presence of carbon in the substrate should be 20–30 times more than that of nitrogen [51].

4.9 The concept of gasification and its types of reactors

The clean combustible producer gas is formed by the gasification of raw materials used for biomass, which results in lower emission of dangerous substances into the ecosystem.

The other advantages of gasification have been listed as follows: a unique feature of a gasifier is that it bifurcates the generation of the gas and combustion of gas, which can be eliminated before burning. It enhances the classification of the wastes from the produced gas in the stream of the gas prior to any utilization. A wide variety of gasifiers are utilized to overcome many needs that lead to state gasification as an improved process. From the monetary point of view, gasification is utilized efficiently on a small scale. Compared to other processes, it is seen that gasification gives rise to efficiency and also provides convenience for the operation. The newly produced gas, as a result of gasification, rise in the span of engines life as it has low ignition temperature with respect to the natural gas [53].

The heat needed to perform for the endothermic reactions can be given by direct means or by indirect means. In gasification, where heat is given directly, the reactions in the process of pyrolysis and the process of gasification conducted in a single vessel. The syngas gasifiers and Gas Technology Institute (GTI) are some of the major examples.

BCL/FERCO gasifier based on indirectly heated gasification technology makes use of a bed of hot particles (sand) that are liquidized along with the usage of steam. Reaction of combustion is separated from the remaining reactions to form a product gas that is free from the nitrogen gas having a calorific value of 15 MJ/m^3 that is 403 Btu/ft^3 [54].

Commercial reactors that are utilized for the process of gasification are mainly classified as shown in Fig. 4.1. Table 4.2 gives different gasifier types performance data.

4.9.1 Fixed bed gasification

The most used gasifier for the last many years that is conventional and indispensable are termed to be FBG or packed bed gasifier. It mainly consists of vessels in a cylindrical shape that is occupied with a pellet-shaped catalyst that is between the range of 1 and 10 mm, which has made it very easy for the operation and designed in simplicity. To use the ceramic beads, which are inert in the state for the distribution of feeds, which are similar to the catalyst bed with the help of catalyst that is metallic supported.

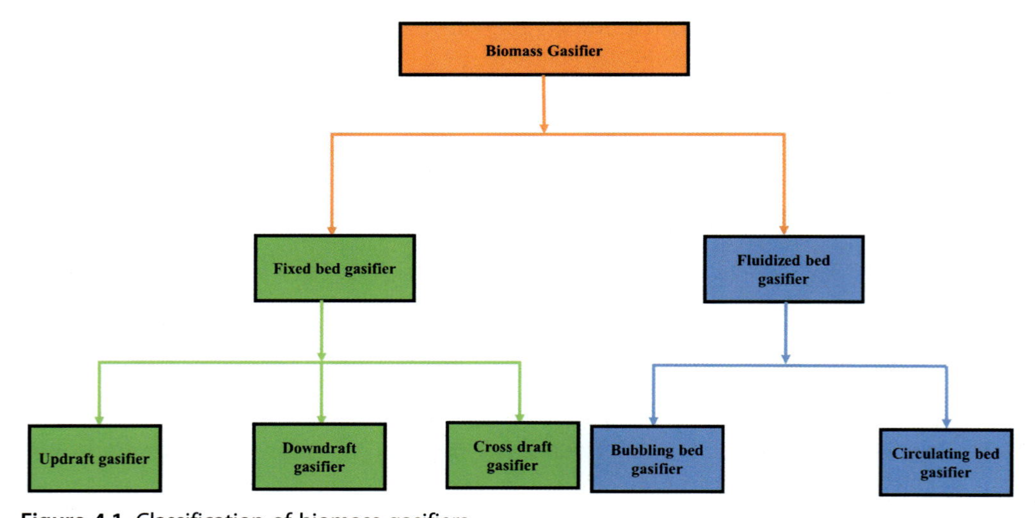

Figure 4.1 Classification of biomass gasifiers.

Table 4.2 Performance data of most commonly used gasifier types [7].

Gasification technology	Gasification temperature (°C)	Char conversion (%)	Cold gas efficiency (%)	Tar content in raw producer gas (g/m^3)
Fixed bed updraft gasifier	Max. bed temp.: 950–1150Gas exit temp.: 150–400	40–85	20–60	30–150
Fixed bed downdraft gasifier	Max. bed temp.: 900–1050Gas exit temp.: 700	<85	30–60	0.015–0.5
Circulating fluidized bed gasifier	750–850	70–95	50–70	5–12
Fluidized bed gasifier	800–900	<70	<70	10–40

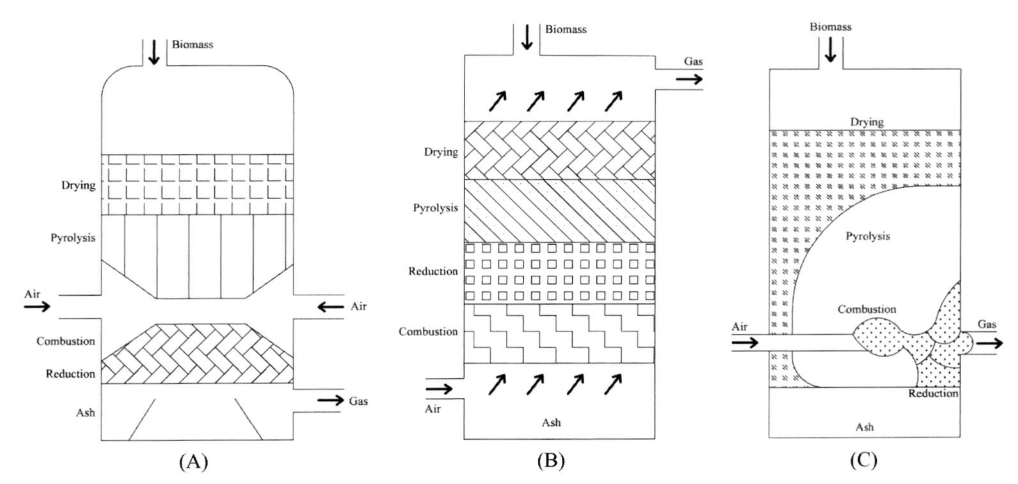

Figure 4.2 (A) Schematic view of Downdraft gasifier, (B) Schematic view of Updraft gasifier, (C) Schematic view of Crossdraft gasifier [56].

FBG consists of many benefits and drawbacks in the process. The benefits are the cheap cost for maintenance, constant behavior in the plug flow, and very little wear and tear. But the drawbacks are a very poor distribution of the heat all over the FBG with the high rate of reactions in the exothermic and endothermic reactions that is very hard for maintaining the temperature. The catalyst may be deactivated due to this environment leading to the end of this process.

Fixed bed reactors are classified into three main types based on the airflow: updraft gasifier, downdraft gasifier, and cross-flow gasifier [55] (Fig. 4.2).

4.9.2 Updraft gasifier

The top of the updraft gasifier is made to put the feedstock, and the atmospheric air is allowed to blow from the base of the updraft gasifier with the help of the grate and is released from the top section. A solid residue that is made from the release of the volatiles called char is placed above the grate and is generated due to rise in the temperature and the combustion to around 1000°C. Also, the hot gases passing upward are dropped, and the ash content moving downward to the base is passed from the grate. The process of pyrolysis takes place for the biomass in the upper section of the gasifier and the drying process at the exit with a temperature of around 200°C–300°C. The volatile HC seen in the vapor leads to a rise in the energy present in the gas, leaving in this temperature. The tar produced is partially allowed to condense in the biomass that exits the pyrolysis zone along with the gas [55].

4.9.3 Downdraft gasifier

Biomass made up of wood can be used as feedstock for downdraft gasification. Here, it is seen that the traveling path of the ash and the feedstock are unidirectional. At the time of gasification, partially burned tar is produced in the hot region through which the producer gas is passed. The syngas produced has a temperature range of 900°C and 1000°C containing some contents of ash and soot. The high temperature seen in the syngas produced leads to lowering of the efficiency of gas formed in the downdraft gasifier when compared to the updraft gasifier [55].

4.9.4 Cross-flow gasifier

A cross-flow gasifier takes place with the utilization of heat sources, which is volatiles, that is partially combusted and charcoal. In the cross-flow gasifier, the feedstock is entered from the top section and travels to the base, and the air is introduced from one end of the gasification plant and exits from the other side situated at the same height. The gasifier has a zone for high temperature for the airflow region, and at the same time, pyrolysis occurs in the top zone of the gasifier. The gas that exits the gasifier has a temperature of nearly 800°C–900°C with high contents of tar that leads to a drop in the efficiency of the energy based on the type of coal taken in the process. FBG can produce a high quantity of tar or char that are not converted and will not be considered for improvements linked to the gasification of the biomass. But it is seen that the FBG can process any inhomogeneous biomass that is the solid waste from the municipal and also has great uses for converting waste to fuels [55].

4.9.5 Fluidized bed gasification

It is seen that in coal gasification, (FLBG) is highly utilized in the last few years. The bed of the fluidized bed behaves as fluid due to its O_2-rich spread over the fine matters on the bed. Back mixing is also done in this gasifier. It is also seen that the velocity of the gas flow is high in the fluidized bed compared to the FBG. It is also seen that the temperature distribution in the gasification phases is constant that is due to the fluidized bed and also does not allow mixing of the matters in the hotbed, biomass, and the hot output gas due to combustion, which leads to no existence of distinguished reaction zone. Hence, it allows all the processes to take place at the same time, such as oxidation, pyrolysis, and drying.

Many sugar mills all over India make use of these power plants that make use of the FLBG method that has a capacity of nearly 5–35 MW. Moreover, bubbling fluidized bed gasification, also known as BFLBG, circulating fluidized bed gasification, also known as CFLBG, double bed gasification, also known as DB, and enriched bed gasification, also known as EB, are the types of FLBG [55] (Fig. 4.3).

4.9.6 Bubbling fluidized bed gasification

It is seen that for the process of gasification of the biomass, the highly shown method for gasification is the BFLB gasification. The BFLB gasifiers give high convenience for doing the operation for many factors such as capacity of the gasifier, the temperature of the process, pressure taken, and also the biomass used. The

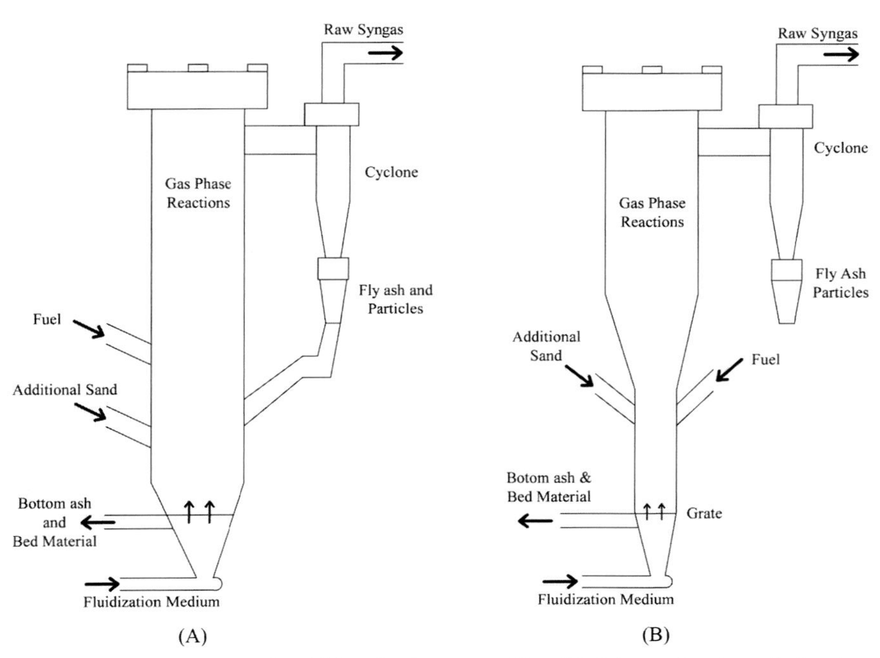

Figure 4.3 (A) Schematic view of Bubbling bed type gasifier, (B) Schematic view of Circulating bed type gasifier [56].

syngas produced is high when the temperature in the gasifier is above 1200°C−1300°C, at the same time, the H_2 and CH_4 contents in the syngas stay low when the temperature is below 1000°C. Hence, the process of BFLB gasification is termed to be the best method for the synthesis along with its utilization. The BFLB process makes use of the particles that are inert in nature, which is sand or alumina of a respective size. The density of the particle and thermal properties of the particles is taken into consideration. The oxidant that is O_2, air, or H_2O steam is forcefully entered with a given velocity that maintains the equilibrium of the mass of the solid, which are known to be minimum fluidization velocity, where the process of bubbling of the gas and channeling takes place in the boiling state and also gives high transfer of heat to the feedstock that is fed to the gasifier facilitating the reaction process [57] (Table 4.3).

4.10 Life cycle analysis of gasification process

4.10.1 Scope of analysis and definition

Environmental consequences of biomass gasification include hazardous element emissions, polyhalogenated aromatic chemicals, heavy metals, and tar production.

4.10.2 Boundary system and analysis of related legislation

The boundary of the system is analyzed, and all components must be defined. All subsystems that have some use or purpose is considered as separate components. Regulatory and policy frameworks create the market pull and market push to develop bioenergy production. On a smaller scale, there is a lack of policies and incentives that support development.

4.10.3 Proper selection of environmental performance indicators

The performance of gasifiers is continuously noted with the help of environmental performance indicators. These indicators simplify and summarize all the information about a complicated system's current condition. The performance of the gasifier is calculated by the selection of appropriate raw data and also the relationship between each one of them.

4.10.4 Inventory analysis

It is the phase that involves the compilation of all the inputs and outputs associated with the biofuel production process using gasifiers and structures the flow of the

Table 4.3 Comparison between different gasifier types [58].

Gasifier type	Advantages	Disadvantages
Fixed bed ("updraft," "downdraft," and "cross draft")	• Strong residence time. • Conversion of fuel. • High thermal efficiency.	• High tar presence in syngas. • Operating and temperature controls are difficult. • Uniform sizes and low biomass humidity are required.
	• Biomass controlling capability with various sizes and high humidity.	• CO and H_2 production is low.
	• Exceptional biomass interaction with the reacting atmosphere.	• Limited feeding and processing versatility (similar properties are must).
	• Achievability of output on a broad scale.	• Difficulty starting and temperature regulation.
Fluidized bed ("circulating" and "bubbling")	• Elevated biomass−atmosphere interaction. • Carbon conversion. • Thermal loads.	• Carbon loss in ashes.
	Easy for controlling of temperature, feeding and refining, and biomass handling with different characteristics.	Difficult in the removal of dust and ash formation
	Preferable for pretreated waste and agricultural waste.	Process temperature is very low; reduction in size.
	Catalyst can be used on a large scale, and the residence time is less.	Construction and maintenance cost are higher.Difficult to work on complex technology; the higher temperature will not influence the biomass particle size.
Entrained flow reactor	Materials can be chosen, and it is easy to use the parameters' process.	Oxidants are needed
	Due to high-temperature sluggish-temperature slagging operation (vitrified slag).	Size and supply of preparation must be reduced
	It can be used on a larger scale.	Efficiency is increased only with heat recovery.
	Low composition of tar and high conversion into carbon.	Low efficiency with cold gas.
	Constant temperature and less residence time in the reactor.	Smaller life, higher cost for investment and maintenance.
Rotary kilns	High capacity and can be used on a large scale. Different biomass can be used and loaded.	The starting temperature, leakage, and wear are hard to control.
	Construction is simple and reliable.	The high content of tar and dust. Heat exchange capacity is less.
	Less investment.	Low efficiency, less flexibility, and high cost for maintenance.
Plasma reactor	Slag, which is inert and not leachable, is produced, which may contain heavy metals.	Nanoparticles found in the syngas lead heat shocking during starting and shut down.
	The waste that is leachable can be used as a building material.	Electrodes and refractory are consumed during the process.
	When using on large scale reaction time will be less in the reactor resulting in lower flow rate of clean syngas.	Safety concerns and molten material must be removed from the ducts.

functional units, namely pyrolysis and gasification in the field of energy, materials, and production output.

4.10.5 Environmental impact assessment

The life cycle impact evaluation determines the amount and magnitude of potential environmental impacts from functional units, as well as appropriate impact classes based on the assessment. A life cycle assessment (LCA) may be performed using a variety of approaches, such as the Tool for the Reduction and Assessment of Chemical and Other Environmental Impacts or the Institute of Environmental Sciences.

4.10.6 Life cycle assessment

When the results are compared to mathematical modeling and experimental data, significant conclusions can be drawn. A sensitivity analysis was carried out to identify the factors that had the greatest influence on the results and to reduce the influence of inaccurate data on the conclusions. The sensitivity analysis variables were chosen to reflect system regions with intrinsically higher unknowns in the data [59].

4.11 Aspen plus approach to the biomass gasification system

Due to its heterogeneous character, the biomass feedstock is classified as a "nonconventional" chemical in aspen plus modeling. As a result, until it is transformed into its usual compounds, it does not take part in calculations of phase or chemical equilibrium (C, Cl_2, H_2O, O_2, S, N_2, H_2, etc.) [60]. Nonconventional constituents are generally represented by their component qualities, which are based on their proximal and final analyses. The type of the property method must be stated in light of the process's requirements [61]. In gasification modelling, the Peng–Robinson cubic equation-of-state (PENG-ROB) technique is the most commonly employed property approach [61], for ideal technique the cubic equation-of-state Redlich–Kwong–Soave (RKS), for alpha function of Boston-Mathias (PR-BM) [62], the cubic equation-of-state RKS and using Boston-Mathias alpha function and RKS (RKS-BM). The built-in models "HCOALGEN" and "DCOALIGT" are used to determine the two major characteristics of biomass, namely density and enthalpy. The creation of a process flowsheet follows the definition of the process requirements. As shown in Fig. 4.4, the process is broken into several subprocesses, each of which is represented by selecting the most relevant operation block. To represent a specific gasification process, use S3 or a combination of the other blocks. Fig. 4.4 shows a flowchart of the procedures involved in biomass gasification [61].

Drying is the initial phase; in this process, a hot drying agent is coupled with biomass, such as flue gas, air, or nitrogen, and delivered to the RStoic block, with the

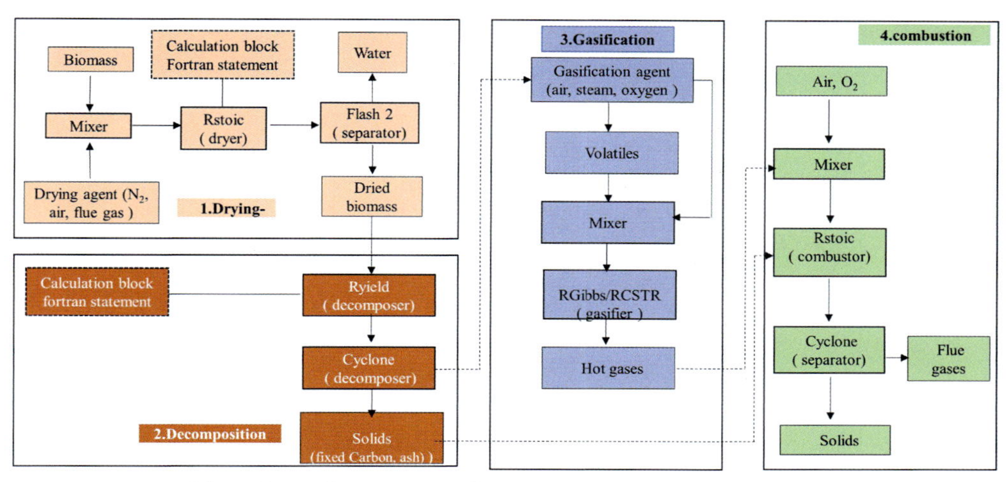

Figure 4.4 General flow sheet of Biomass gasification in Aspen Plus [61].

assumption that the process of removing moisture from biomass is a chemical reaction [63]. It is worth noting that drying takes place in the RStoic block usually refers to the "drying phase" that occurs during the gasification process. Depending on the moisture content of the biomass feedstock, it may be necessary to dry the biomass beforehand as a "pretreatment step." In this circumstance, other dryer blocks, such as a convective dryer or a contact drier, can be utilized to simulate this step. More design details, such as length, flow direction, particle size, residence time, or drying kinetics, are all available and may be required for these dryer models. In the RYield block, drying is followed by breakdown to transform biomass into its "conventional" components. It is anticipated that during biomass decomposition, the ultimate and proximate analyses of the fuel can provide information on the yield and composition of the devolatilization products. Following decomposition, biomass gasification occurs, which will be discussed in further depth later. The RStoic block is used to mimic incomplete combustion as the last stage. To be combusted, char is combined with oxygen or air and then passed through a cyclone block, which separates the final gas products from the solids. Depending on the goal, design, or complexity of the overall process, these phases may include additional blocks or user-defined computation blocks. Modeling of biomass gasification may be divided into two categories: equilibrium and kinetic. The equilibrium technique, which may be classed as stoichiometric or nonstoichiometric, is used to forecast by assuming that the reactants are completely mixed and will continue to react with each other forever; the syngas composition can be calculated. Stoichiometric models rely on equilibrium constants and need the selection of relevant reactions in addition to the equilibrium constants to estimate the final product composition. However, due to a lack of relevant information, this strategy may result in some inaccuracies, especially for complicated systems. As a result, a nonstoichiometric modeling technique based on Gibbs free energy minimization can be used to solve this problem. One of the benefits of this method is that it requires very little system specifics for the process description and result prediction.

In aspen plus, equilibrium modeling is done with the RGibbs block, which is based on minimizing the system's Gibbs free energy in terms of the mole numbers of the species present in all phases, such as reactants and products. This strategy specifies may be found elsewhere. Kinetic modeling, on the other hand, is commonly used to estimate the syngas production and composition for a particular operating state and gasifier design after a defined period (or in a defined volume in a fluid). As a result, this technique is more realistic in terms of the actual condition in the gasifier, especially at lower gasification temperatures and shorter residence times. Gasification reactions are often utilized in kinetic modeling. A user-defined Fortran code in aspen plus can be used to integrate the reaction kinetics. The reaction kinetics, as well as the gasifier's bed hydrodynamics, can be incorporated into this technique. The majority of research use complete kinetic models to represent biomass gasification, while semikinetic and reversible rate expressions were used less frequently. Semikinetic models were most typically used while modeling using aspen plus. RCSTR and RPLUG are two process blocks that can be employed. The RCSTR block is often used to investigate various types of fluidized-bed reactors to replicate conversion behavior due to its perfect mixing assumption. It is feasible to model the reactor bed and free-bed sections as two connected RCSTR blocks that incorporate the hydrodynamic bed characteristics. RPLUG, on the other hand, allows for the investigation of many configurational characteristics, such as the gasifier's ideal length, height, and diameter. As a result, selecting the appropriate aspen plus block has a significant impact on the process's progress [61].

4.12 Conclusion

Biomass can be considered an appropriate alternative to fossil fuels. As mentioned in this chapter, technologies, such as gasification and their variants, are effective in converting the energy content of biomass into valuable products. Researchers are still looking to reduce the technology's limits and improve the energy efficiency of the producing gas. This chapter highlighted the TCC technologies, production and use of aquatic biomass, lignocellulose biomass pretreatment, bioconversion and downstream processing biomass-derived molecules' conversion to chemicals, types of gasification reactors, and LCA and aspen plus approach to gasification system. Future work might include investigating the impact of various catalysts on the gasification process. IGCC is a promising technology that may be used to make use of the abundant biomass resources while diversifying electricity generation. Accurate assessment of ash, tar, suspended particle matter, and moisture in producer gas is a potential technological problem that can aid in the development of recommendations for the use of producer gas.

References

[1] Devi GS, Vaishnavi S, Srinath S, Dutt B, Rajmohan KS. Energy recovery from biomass using gasification. Curr Dev Biotechnol Bioeng Resour Recover Wastes 2020. Available from: https://doi.org/10.1016/B978-0-444-64321-6.00019-7.

[2] BP. BP Statistical Review of World Energy, 67th ed. Statistical Review of World Energy; 2018.

[3] Zhang W. Automotive fuels from biomass via gasification. Fuel Process Technol 2010. Available from: https://doi.org/10.1016/j.fuproc.2009.07.010.

[4] Basu P. Combustion and gasification in fluidized beds. <https://doi.org/10.1201/9781420005158>; 2006.

[5] Chen GL, Chen GB, Li YH, Wu WT. A study of thermal pyrolysis for castor meal using the Taguchi method. Energy 2014. Available from: https://doi.org/10.1016/j.energy.2014.04.009.

[6] Hu X, Gholizadeh M. Biomass pyrolysis: a review of the process development and challenges from initial researches up to the commercialisation stage. J Energy Chem 2019. Available from: https://doi.org/10.1016/j.jechem.2019.01.024.

[7] Heidenreich S, Foscolo PU. New concepts in biomass gasification. Prog Energy Combust Sci 2015. Available from: https://doi.org/10.1016/j.pecs.2014.06.002.

[8] Heidenreich S, Müller M, Foscolo PU. Advanced biomass gasification: new concepts for efficiency increase and product flexibility. <https://doi.org/10.1016/C2015-0-01777-4>; 2016.

[9] Chynoweth DP, Owens JM, Legrand R. Renewable methane from anaerobic digestion of biomass. Renew Energy 2001. Available from: https://doi.org/10.1016/S0960-1481(00)00019-7.

[10] Sawatdeenarunat C, Surendra KC, Takara D, Oechsner H, Khanal SK. Anaerobic digestion of lignocellulosic biomass: challenges and opportunities. Bioresour Technol 2015178:178–86. Available from: https://doi.org/10.1016/j.biortech.2014.09.103.

[11] Perea-Moreno MA, Samerón-Manzano E, Perea-Moreno AJ. Biomass as renewable energy: worldwide research trends. Sustainability 201911. Available from: https://doi.org/10.3390/su11030863.

[12] Cheng JJ. Biomass to renewable energy processes, 2nd ed. <https://doi.org/10.1201/9781315152868>; 2017.

[13] Cho EJ, Trinh LTP, Song Y, Lee YG, Bae HJ. Bioconversion of biomass waste into high value chemicals. Bioresour Technol 2020. Available from: https://doi.org/10.1016/j.biortech.2019.122386.

[14] Wunderlich SM, Martinez NM. Conserving natural resources through food loss reduction: production and consumption stages of the food supply chain. Int Soil Water Conserv Res 2018. Available from: https://doi.org/10.1016/j.iswcr.2018.06.002.

[15] Singh CK, Kumar A, Roy SS. Quantitative analysis of the methane gas emissions from municipal solid waste in India. Sci Rep 2018. Available from: https://doi.org/10.1038/s41598-018-21326-9.

[16] Sarkar N, Ghosh SK, Bannerjee S, Aikat K. Bioethanol production from agricultural wastes: an overview. Renew Energy 2012. Available from: https://doi.org/10.1016/j.renene.2011.06.045.

[17] Koutinas AA, Vlysidis A, Pleissner D, Kopsahelis N, Lopez Garcia I, Kookos IK, et al. Valorization of industrial waste and by-product streams via fermentation for the production of chemicals and biopolymers. Chem Soc Rev 2014. Available from: https://doi.org/10.1039/c3cs60293a.

[18] Djilas S, Čanadanović-Brunet J, Ćetković G. By-products of fruits processing as a source of phytochemicals. Chem Ind Chem Eng Q 2009. Available from: https://doi.org/10.2298/CICEQ0904191D.

[19] Padam BS, Tin HS, Chye FY, Abdullah MI. Banana by-products: an under-utilized renewable food biomass with great potential. J Food Sci Technol 2014. Available from: https://doi.org/10.1007/s13197-012-0861-2.

[20] Caputo AC, Scacchia F, Pelagagge PM. Disposal of by-products in olive oil industry: waste-to-energy solutions. Appl Therm Eng 2003. Available from: https://doi.org/10.1016/S1359-4311(02)00173-4.
[21] Kondamudi N, Mohapatra SK, Misra M. Spent coffee grounds as a versatile source of green energy. J Agric Food Chem 2008. Available from: https://doi.org/10.1021/jf802487s.
[22] Nawirska A, Kwaśniewska M. Dietary fibre fractions from fruit and vegetable processing waste. Food Chem 2005. Available from: https://doi.org/10.1016/j.foodchem.2003.10.005.
[23] Nguyen QA, Cho E, Trinh LTP, Jeong J-S, Bae H-J. Development of an integrated process to produce D-mannose and bioethanol from coffee residue waste. Bioresour Technol 2017. Available from: https://doi.org/10.1016/j.biortech.2017.07.169.
[24] Babbar N, Dejonghe W, Gatti M, Sforza S, Elst K. Pectic oligosaccharides from agricultural by-products: production, characterization and health benefits. Crit Rev Biotechnol 2016. Available from: https://doi.org/10.3109/07388551.2014.996732.
[25] Kumar R, Strezov V, Weldekidan H, He J, Singh S, Kan T, et al. Lignocellulose biomass pyrolysis for bio-oil production: a review of biomass pre-treatment methods for production of drop-in fuels. Renew Sustain Energy Rev 2020. Available from: https://doi.org/10.1016/j.rser.2020.109763.
[26] Muthudineshkumar R, Anand R. Anaerobic digestion of various feedstocks for second-generation biofuel production. Advances in eco-fuels for a sustainable environment. 2019. Available from: https://doi.org/10.1016/b978-0-08-102728-8.00006-1.
[27] Moset V, Cerisuelo A, Sutaryo S, Møller HB. Process performance of anaerobic co-digestion of raw and acidified pig slurry. Water Res 2012;46(16):5019–27. Available from: https://doi.org/10.1016/j.watres.2012.06.032.
[28] Braguglia CM, Gagliano MC, Rossetti S. High frequency ultrasound pretreatment for sludge anaerobic digestion: effect on floc structure and microbial population. Bioresour Technol 2012. Available from: https://doi.org/10.1016/j.biortech.2012.01.074.
[29] Agrahari R, Tiwari GN. The production of biogas using kitchen waste. Int J Energy Sci 2013. Available from: https://doi.org/10.14355/ijes.2013.0306.05.
[30] Taherzadeh MJ, Karimi K. Pretreatment of lignocellulosic wastes to improve ethanol and biogas production: a review. Int J Mol Sci 2008. Available from: https://doi.org/10.3390/ijms9091621.
[31] Ariunbaatar J, Panico A, Esposito G, Pirozzi F, Lens PNL. Pretreatment methods to enhance anaerobic digestion of organic solid waste. Appl Energy 2014. Available from: https://doi.org/10.1016/j.apenergy.2014.02.035.
[32] Mussoline W, Esposito G, Giordano A, Lens P. The anaerobic digestion of rice straw: a review. Crit Rev Env Sci Technol 2013. Available from: https://doi.org/10.1080/10643389.2011.627018.
[33] Duan N, Dong B, Wu B, Dai X. High-solid anaerobic digestion of sewage sludge under mesophilic conditions: feasibility study. Bioresour Technol 2012. Available from: https://doi.org/10.1016/j.biortech.2011.10.090.
[34] Sambusiti C, Ficara E, Malpei F, Steyer JP, Carrère H. Benefit of sodium hydroxide pretreatment of ensiled sorghum forage on the anaerobic reactor stability and methane production. Bioresour Technol 2013. Available from: https://doi.org/10.1016/j.biortech.2013.06.095.
[35] Teymouri F, Laureano-Perez L, Alizadeh H, Dale BE. Optimization of the ammonia fiber explosion (AFEX) treatment parameters for enzymatic hydrolysis of corn stover. Bioresour Technol 2005. Available from: https://doi.org/10.1016/j.biortech.2005.01.016.
[36] Saritha M, Arora A, Lata. Biological pretreatment of lignocellulosic substrates for enhanced delignification and enzymatic digestibility. Indian J Microbiol 2012. Available from: https://doi.org/10.1007/s12088-011-0199-x.
[37] Cheng JJ, Timilsina GR. Status and barriers of advanced biofuel technologies: a review. Renew Energy 2011. Available from: https://doi.org/10.1016/j.renene.2011.04.031.
[38] Bhutto AW, Qureshi K, Harijan K, Abro R, Abbas T, Bazmi AA, et al. Insight into progress in pre-treatment of lignocellulosic biomass. Energy 2017. Available from: https://doi.org/10.1016/j.energy.2017.01.005.
[39] Sheldon RA. Green and sustainable manufacture of chemicals from biomass: state of the art. Green Chem 2014. Available from: https://doi.org/10.1039/c3gc41935e.

[40] Liu WJ, Jiang H, Yu HQ. Development of biochar-based functional materials: toward a sustainable platform carbon material. Chem Rev 2015. Available from: https://doi.org/10.1021/acs.chemrev.5b00195.
[41] Wang S, Dai G, Yang H, Luo Z. Lignocellulosic biomass pyrolysis mechanism: a state-of-the-art review. Prog Energy Combust Sci 2017. Available from: https://doi.org/10.1016/j.pecs.2017.05.004.
[42] Fatih Demirbas M. Current technologies for biomass conversion into chemicals and fuels. Energy Sources A Recovery Util Environ Effects 2006. Available from: https://doi.org/10.1080/00908310500434556.
[43] Lee J, Yang X, Cho SH, Kim JK, Lee SS, Tsang DCW, et al. Pyrolysis process of agricultural waste using CO_2 for waste management, energy recovery, and biochar fabrication. Appl Energy 2017. Available from: https://doi.org/10.1016/j.apenergy.2016.10.092.
[44] Sun Y, Yu IKM, Tsang DCW, Cao X, Lin D, Wang L, et al. Multifunctional iron-biochar composites for the removal of potentially toxic elements, inherent cations, and hetero-chloride from hydraulic fracturing wastewater. Environ Int 2019. Available from: https://doi.org/10.1016/j.envint.2019.01.047.
[45] Quaak P, Knoef H, Sfassen H. Energy from biomass. A review of combustion and gasification technologies. World Bank Technical Paper; 1999.
[46] Belgiorno V, De Feo G, Della Rocca C, Napoli RMA. Energy from gasification of solid wastes. Waste Manag 2003. Available from: https://doi.org/10.1016/S0956-053X(02)00149-6.
[47] Bridgwater AV. Catalysis in thermal biomass conversion. Appl Catal A Gen 1994. Available from: https://doi.org/10.1016/0926-860X(94)80278-5.
[48] Najjar YSH. Comparison of performance of the integrated gas and steam cycle (IGSC) with the combined cycle (CC). Appl Thermal Eng 1999. Available from: https://doi.org/10.1016/s1359-4311(98)00008-8.
[49] Heppenstall T. Advanced gas turbine cycles for power generation: a critical review. Appl Thermal Eng 1998. Available from: https://doi.org/10.1016/S1359-4311(97)00116-6.
[50] Rao PV, Baral SS, Dey R, Mutnuri S. Biogas generation potential by anaerobic digestion for sustainable energy development in India. Renew Sustain Energy Rev 2010. Available from: https://doi.org/10.1016/j.rser.2010.03.031.
[51] Deepanraj B, Sivasubramanian V, Jayaraj S. Biogas generation through anaerobic digestion process—an overview. Res J Chem Environ 201418.
[52] Nsair A, Cinar SO, Alassali A, Qdais HA, Kuchta K. Operational parameters of biogas plants: a review and evaluation study. Energies 2020. Available from: https://doi.org/10.3390/en13153761.
[53] Bartela è, Kotowicz J, Dubiel-Jurgaś K. Investment risk for biomass integrated gasification combined heat and power unit with an internal combustion engine and a Stirling engine. Energy 2018. Available from: https://doi.org/10.1016/j.energy.2018.02.152.
[54] Biomass integrated gasifier combined cycle technology: III. Application in the cane sugar industry. Fuel Energy Abstr. <https://doi.org/10.1016/s0140-6701(01)80416-1>; 2001.
[55] Sansaniwal SK, Rosen MA, Tyagi SK. Global challenges in the sustainable development of biomass gasification: an overview. Renew Sustain Energy Rev 2017. Available from: https://doi.org/10.1016/j.rser.2017.05.215.
[56] Sansaniwal SK, Pal K, Rosen MA, Tyagi SK. Recent advances in the development of biomass gasification technology: a comprehensive review. Renew Sustain Energy Rev 2017. Available from: https://doi.org/10.1016/j.rser.2017.01.038.
[57] Ul Hai I, Sher F, Yaqoob A, Liu H. Assessment of biomass energy potential for SRC willow woodchips in a pilot scale bubbling fluidized bed gasifier. Fuel 2019. Available from: https://doi.org/10.1016/j.fuel.2019.116143.
[58] Ren J, Cao JP, Zhao XY, Yang FL, Wei XY. Recent advances in syngas production from biomass catalytic gasification: a critical review on reactors, catalysts, catalytic mechanisms and mathematical models. Renew Sustain Energy Rev 2019. Available from: https://doi.org/10.1016/j.rser.2019.109426.

[59] Zang G, Zhang J, Jia J, Lora ES, Ratner A. Life cycle assessment of power-generation systems based on biomass integrated gasification combined cycles. Renew Energy 2020. Available from: https://doi.org/10.1016/j.renene.2019.12.013.
[60] Al-Malah KIM. Aspen plus: chemical engineering applications. J Chem Inf Model 2017.
[61] Mutlu ÖÇ, Zeng T. Challenges and opportunities of modeling biomass gasification in aspen plus: a review. Chem Eng Technol 2020. Available from: https://doi.org/10.1002/ceat.202000068.
[62] Doherty W, Reynolds A, Kennedy D. Aspen plus simulation of biomass gasification in a steam blown dual fluidised bed. Materials and processes for energy: communicating current research and technological developments. Formatex Research Centre 2013.
[63] Aspen Technology Inc. Getting started modeling processes with solids. Aspen Technol Inc.; 2013.

CHAPTER 5

Life Cycle Assessment applied to waste-to-energy technologies

5.1 Introduction

The fight against climate change requires the adoption of a series of measures to avoid the emission of greenhouse gases (GHG), with increased participation of renewable sources in the world's energy matrix as noteworthy. In this sense, biofuels are considered an adequate alternative to replace fossil fuels both in electricity generation and in the transportation of cargo and passengers.

Biofuels can be obtained in different ways, such as through the fermentation process, in the case of ethanol, or by chemical extraction, used for seeds and grains, like soybean, or by incineration, gasification and the pyrolysis of biomass residues. These latest technologies, also called waste-to-energy (WtE), have gained importance as alternatives for renewable energy sources. Through their application, waste can be converted into heat, electricity, and liquid or gaseous fuels. Incineration is the simplest and the most widely applied process, based on waste combustion to generate heat, whereas gasification and pyrolysis comprise more complex steps, but also allow for other forms of energy to be generated, and are in full development.

In addition to cleaner energy generation, WtE technologies also consist in socio-environmental benefits related to the avoidance of the inappropriate disposal of waste, which is often disposed of in landfills or even in open dumps. Thus the energy recovery of waste can contribute not only to mitigate climate change by replacing the use of fossil fuels but also to more broadly improve environmental quality since it allows for more adequate waste management.

WtE technologies also allow for small-scale energy production, which can be an interesting solution for rural communities. Gasification and pyrolysis products can be used for residential heating and food preparation, or for generating electricity. In this way, these technologies can also fulfill the function of promoting energy access for the less economically favored population, who often use noncommercial sources of energy, such as firewood, animal waste, and the biomass waste itself.

However, despite these benefits, WtE technologies can also result in environmental impacts, in addition to displaying certain limitations in their use. Impurities present in the gas, both organic (tar) and inorganic (H_2S, HCl, NH_3, HCN and alkali metals), pose challenges for their use and can raise costs [1]. In this sense, it is

A Thermo-Economic Approach to Energy From Waste
DOI: https://doi.org/10.1016/B978-0-12-824357-2.00004-8

essential to quantify the benefits of the aforementioned technologies, considering their positive and negative aspects. In this regard, the life cycle assessment (LCA) has proved to be a useful tool for this analysis, with increasing applications noted in the literature.

The purpose of this chapter is to present the LCA methodology and its applications in the environmental analysis of WtE technologies. The first part is dedicated to the LCA itself, addressing its development history, its applications, and the main steps that constitute an LCA study. This part is intended for readers with little or no knowledge of the methodology and aims to provide basic understanding. The second part of the chapter is dedicated to the application of the LCA to WtE technologies, presenting the main applications in this regard, the elements highlighted in the analysis, and the main results reported in the literature. The main technologies that are analyzed in LCA studies are also briefly presented. This chapter can be useful for readers who have basic knowledge of both LCA and WtE technologies.

5.2 What is life cycle assessment?

Many definitions of what LCA is are found in books, reports, and articles, varying according to the author's vision or writing style, but in general they should all contain certain fundamental terms, such as "impact assessment," "flow (input and output) of energy and matter," and "cradle to the grave." These three elements are essential precisely because they manage to approximate the idea behind the LCA, which consists of an assessment (or comparison) of impacts, considering the analysis of all energy and material flows (including the use of natural resources and emissions to air, water, and soil) that occur from the extraction of raw materials to their final disposal, going through the production processes and their stage of use therefore from the "cradle to the grave."

An important step in understanding LCA is to understand life cycle thinking (LCT). Understanding the LCT we can see that the life cycle of a product or service begins with the extraction of natural resources and the generation of energy. These are then used in production, storage, distribution, use, and maintenance. After the use stage, the products can be reused, recycled, recovered, or sent for final disposal. This flow is represented in Fig. 5.1.

Going a step further, LCA uses this thinking to, as the name suggests, assess the sustainability of products, services, and processes, seeking to be as comprehensive as possible. For this, the LCA comprises steps to define their objective and scope, inventory analysis, inventory evaluation, and interpretation, which will be addressed later in this chapter. Therefore while the LCT is a model that seeks to structure the stages of the life cycle of a product or service, the LCA is a tool based on this model and used to assess their sustainability.

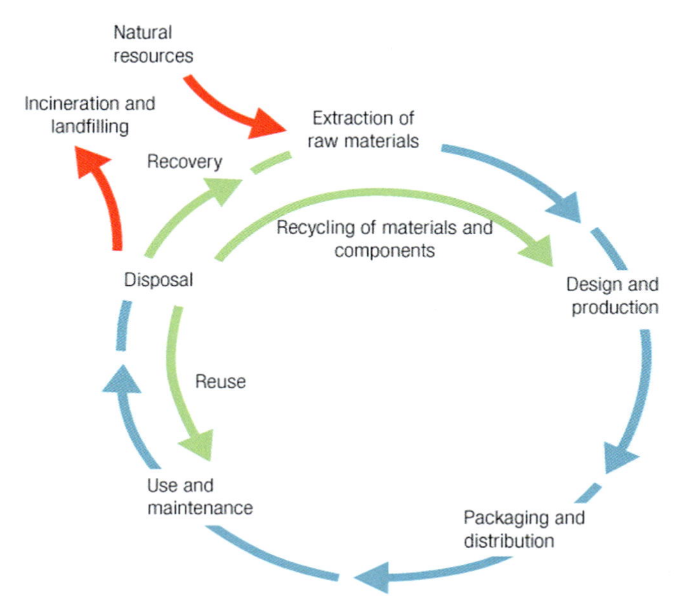

Figure 5.1 Life cycle thinking.

5.2.1 Historical development

The first studies that approach what is defined today as an LCA were carried out from the end of the 1960s. This decade marks the beginning of the growing concern of society with problems related to the environment, such as the diverse forms of pollution (atmospheric, water, and terrestrial), global warming, and depletion of natural resources. In 1962 the publication of the book "Silent Spring," by Rachel Carlson, drew attention to the effects of the use of pesticides on the environment, especially on birds (hence the title of the book), and contributed to the ban of the pesticide DDT in some locations. A decade later, in 1972 the publication of the book *Limits to Growth* explained the contradictions of the current development model, of the unrestrained use of natural resources, arguing that the depletion of these resources would impose a barrier to economic and population growth. In the same year, the Stockholm Conference was held, which was the first meeting of heads of state organized by the United Nations to debate environmental issues and is considered a landmark for the changes in relations between society and the environment.

In this first moment, LCA studies exhibited a more comparative emphasis, seeking to answer whether product A was better than product B in environmental terms. Some practical examples can be found on a daily basis. Is it better to use glass cups, which require water to be washed, or plastic cups, which depend on oil extraction and are discarded after use? Which milk packaging generates less impact: glass, plastic or paper? When choosing between incandescent and fluorescent lamps, the latter consume less energy and last longer, but use more materials and contain heavy metals [2]. A famous study was commissioned by Coca-Cola in 1969 to evaluate two types of soft drink packaging in terms of resource utilization and emissions. The Midwest Research Institute, responsible for the assessment, subsequently conducted a similar study for the United States Environmental Protection Agency and adopted the term Resource and Environmental Profile Analysis [2].

A stagnant period was noted during the 1970s and early 1980s, with little interest in this new tool, little scientific debate and some unpublished studies. This initial LCA development period was characterized by very heterogeneous approaches and terminologies, without a methodological standard, which resulted in very different results, even when the analyzed systems were the same, making it difficult to propagate the tool. This outlook began to change in 1984, when the Swiss Federal Laboratories for Materials Testing and Research published a report that presented a list of data required for LCA studies and introduced a first methodology for assessing life cycle impacts [2].

The 1990s marked the beginning of the development of a standardization for LCA studies, essential for the dissemination of this tool. A first methodological guide was published by the Society for Environmental Toxicology and Chemistry (SETAC) in 1993 after consultation with the international scientific community and was called the Code of Practice. Subsequently, the International Organization for Standardization (ISO) did the same, publishing its first standards in 1997 and 1998, containing methods and procedures for conducting an LCA study. In that decade the first methods for assessing life cycle impacts also appeared, that is, Eco-indicator 95, Eco-indicator 99, CmL 1992, and EDIP/UmIP 97.

The beginning of the 21st century saw major advances in LCA, with new methodological developments, the expansion of application areas and the emergence of organizations dedicated to the theme. New types of products and processes began to be applied as LCA study targets, such as agricultural products and technologies for the use of waste energy. Organizations, such as the Life Cycle Initiative and the European Platform on Life Cycle Assessment, were created in the early 2000s to dedicate themselves to the topic. This development has materialized in an increasing number of scientific articles on LCA published since the 1990s, as well as reports and methodological guides.

5.2.2 Applications of LCA

Currently, the LCA is used in a wide range of applications, ranging from the public sector to civil society and consumers, obviously through the productive and industrial sector. This wide reach of the tool is due to several efforts made to spread its use, carried out by institutions, such as the ISO, the United Nations Environment Program, and the SETAC.

On the public sector side, LCA can be used both for the formulation of new policies and for the implementation and evaluation of existing policies. For example, public policy makers can rely on the LCA to encourage the reduction of impacts in a sector or for specific products, which can occur both through market instruments, such as subsidies, taxation, and labeling, as well as by command-and-control instruments, such as energy efficiency rules. An example of an LCA-based policy is the Integrated Product Policy, developed by the European Commission. However, a recurring challenge for LCA area is noted concerning the possibility of comparing the results of different studies according to the compatibility between the applied methodologies. To overcome this obstacle, it is important that the policy explicitly establish the procedure that must be adopted.

Still with regard to the public sector, LCA can also be used to assess the impacts of policies that already exist or are under development. For example, a government that intends to reduce the fleet of vehicles that use fossil fuels should assess whether the best option, given its context, would be to encourage vehicles powered by biofuels or electricity, and the LCA would play a key role. Another example is waste management. In Sweden, after the introduction of a tax on waste incineration (with the aim of reducing waste generation and increasing recycling and biological treatment), an LCA study indicated that the way the tax was defined encouraged recycling, but that the environmental benefits were small, which led to a reformulation to include the waste carbon content.

From the perspective of the industrial sector, LCA can be classified into five objectives [3]:

1. Decision support in the development of products and processes;
2. Marketing (i.e., eco-labeling);
3. Development of indicators to monitor the environmental performance of products or factories;
4. Selection of suppliers or subcontractors; and
5. Strategic planning.

However, it is common for the same study to serve more than one of the above objectives, and, as the use of LCA is incorporated into company practices, one application can lead to another (i.e., results on the environmental performance of a product can lead to decisions on the choice of suppliers). The use of LCA in the productive sector can also be categorized into two levels: product and corporate.

At the product level, companies often use LCA for product development, to identify aspects to be improved both in internal processes and in the supply chain, and also for environmental labeling. These applications are not exclusive and can actually complement each other. Companies that carry out an assessment to identify product life cycle stages with the greatest impact may end up promoting changes in their design and use the results of the improvement in environmental performance to promote marketing, reaching an increasing group of consumers with environmental concerns.

At the corporate level, it can be used to establish strategic objectives, such as reducing the company's ecological footprint by 20%, or implementing an environmental management system (EMS). In these cases, the use of LCA helps to ensure that the company is not transferring environmental impacts to another stage within the production chain. There is an increasing demand for the application of LCA in EMS.

Finally, from a societal perspective, people as consumers may be exposed to various information related to LCA, from product packaging labels to advertisements from environmentally responsible companies. Consumer decisions based on the LCA may occur, for example, by choosing between a product with less impact over a similar one, by deciding between two different ways to fulfill a function, such as between washing dishes by hand or using a dishwasher, or even the adoption of a less impactful lifestyle (reducing meat consumption, for example) [4].

5.2.3 Steps and procedures for an LCA study

There is no single method for performing an LCA, although current standards and guidelines determine the general procedures for conducting a study based on the specific application and requirements of each user. The ISO 14.040/2001 standard establishes the steps and procedures presented below and illustrated in Fig. 5.2.

5.2.4 Definition of the objective and scope

The purpose and scope of an LCA are closely related and generally presented together. For the objective to be well defined it is important that the scope also be well defined. The definition

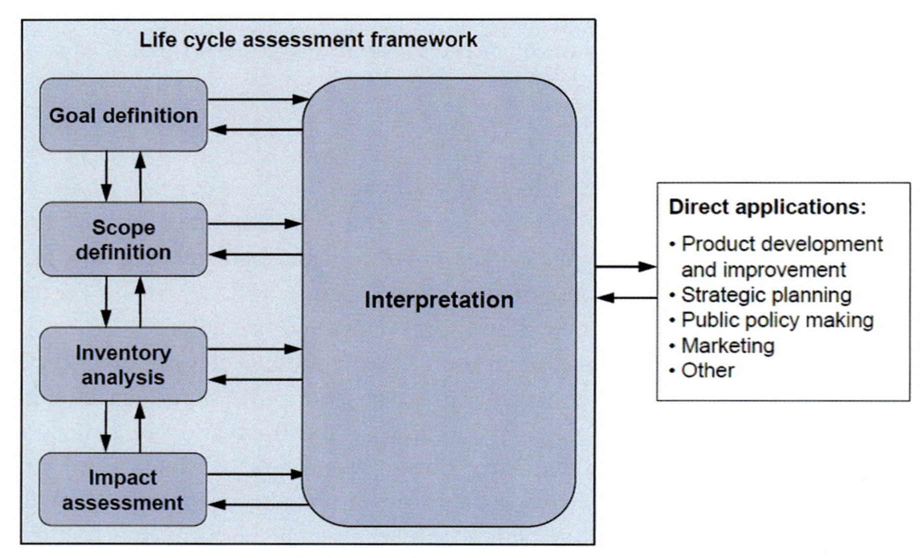

Figure 5.2 Stages of an LCA study. *Source: ISO 14.040/2001.*

of the objective of an LCA should include the intended application for the results, the reasons for conducting the study, the target audience, who the results of the study should be communicated to, and the intention or not of making comparisons that will be made public.

Regarding the intended application, it can be considered, for example, that the objective of the study is to compare two alternatives for the disposal of solid waste in a region, which could be recycling (option 1) or incineration (option 2). The reason for conducting the study could be to support the formulation of public policies for adequate waste disposal. The target audience in this case would be policy makers and other actors working with solid waste management.

With regard to the possibility of making public comparisons, special care must be taken due to the impacts that the results can generate. The assertion that the product of a certain brand is better (has less impact) than that of another can have great repercussions, even causing serious losses to the latter.

5.2.5 Analysis of the life cycle inventory

The analysis or construction of the life cycle inventory (LCI) consists of surveying all inputs and outputs of matter and energy throughout the life cycle considering a functional unit (FU). This survey is usually the most arduous stage of an LCA study, especially in situations of data absence. In the inventory analysis, the system boundaries, reference flows and cut-off and allocation problems must be taken into account [5].

For example, a study of biodiesel LCA and some data that must be collected include fuel consumption and emissions during biomass collection, electricity demand and emissions during processing, as well as the waste from the entire process. The reference flows refer to the quantities of matter that pass from one stage to the next in order to reach the FU at the end.

Like all LCA stages and as illustrated in the figure, an inventory analysis is an iterative process. As the study progresses, new issues can be identified, or changes can be made to the objective and scope that lead to changes in data collection procedures so that the objectives are achieved. Different methods for conducting an inventory are available, but this is beyond the scope of this chapter. Readers interested in learning more about these methods can consult Castelo Branco [5].

5.2.6 Life cycle impact assessment

The life cycle impact assessment (LCIA) stage consists of translating the data obtained during the impact inventory stage. In the inventory the data are presented in physical units (eg mass of CO_2 emitted to the atmosphere or volume of water extracted from water bodies). These data do not represent an impact, and for that connection to be made, the LCI step is necessary, which aims to aggregate the information obtained from the inventory in different impact categories to allow for result interpretation.

Some LCA studies do not require this assessment and may not perform it, but this must be in accordance with the objective and scope. For example, Castelo Branco [5] conducted a study whose objective was to assess the real benefit of using a CO_2 capture and storage technology in a coal-fired power plant, with an inventory being sufficient to this end.

An LCIA can comprise four steps, although not all are mandatory:

1. Selection of impact categories,
2. Categorization,
3. Standardization, and
4. Weighting.

The first, as the name suggests, consists of deciding the categories that will be adopted in the study, among several possible ones, such as global warming potential, acidification, eutrophication, and resource depletion, among others. One can also choose to use more aggregated categories, such as human health, the environment, and natural resources. In the first case, the categories are called midpoint, while in the second they are called endpoint. This selection of impact categories is mandatory in an LCIA and must be consistent with the objectives of the study.

As well as the choice of categories, characterization is mandatory in an LCIA and it is at this stage that the inventory data are included in each of the impact categories, using "characterization factors." For example, if the "global warming potential" category is one of those selected, emissions of different gases must contribute to this impact category (CO_2, CH_4, N_2O, etc.), each one differently. Therefore this step is applied to make associations between the flows of matter and energy of the system and the impact categories.

The standardization step is not mandatory in an LCIA, but it is useful for interpretation and comparison between different studies. Normalization consists of dividing the values obtained for each impact category by a denominator (specific for each category), called the "normalization base," which allows the result to be made relative to a common reference. These denominators can be, for example, results of total impact or damage of the year for that category in a country, region or continent, or globally.

Weighting, as the name implies, means the allocation of weights for the different impact categories, aiming at reaching a single impact value. The weighting factors (weights) aim to reflect the relative importance of each impact category and are therefore subjective. Weighting can be useful when two alternatives (e.g., products) are to be compared and, in this case, the assigned weights must be the same in both cases.

5.2.7 Interpretation

Although it is presented as the last step in an LCA study, it is important to understand that interpretation is present throughout the iterative process, as illustrated in Fig. 5.2.

Any intermediate results obtained throughout the LCIA and the LCIA can be interpreted, allowing for the identification of necessary changes in the methods used to meet the objective of the study, or even leading to objective and scope adjustments.

5.3 Use of LCA to analyze waste-to-energy technologies

Waste recovery (and, more specifically, its energy use) represents the fourth step in the waste management hierarchy, preceded by reduction, reuse, and recycling, in order of priority. Recovery is also succeeded by the final disposal stage, considered the last option for proper waste management. The fact that it is the penultimate step within the hierarchy may lead one to believe that energy recovery has low environmental sustainability. However, there are cases in which this step is more recommended, for example, when the waste contains hazardous materials, presents different types of materials (heterogeneous), or when the other steps are not viable from an ecological or economic point of view [6].

As well as the literature on LCA in general, the studies that apply this tool for the analysis of energy systems have increased significantly. In the last two decades, over 1000 scientific articles focused on LCA have been published, which can be categorized as "Energy and fuels" [7]. Within this field, the application of the LCA for the analysis of waste management and technologies for the energy use of waste (WtE technologies) has become increasingly common.

5.3.1 Main applications

Several technologies for the use of waste energy are available, and it is not the objective of this chapter to analyze them in depth. The following will briefly present some of the main technologies in order to highlight elements of each one that are important for an LCA.

- Incineration—consists of burning the waste to use the thermal energy. As it is a simpler process, incineration imposes few requirements regarding waste characteristics when compared to other technologies. Although it is simple and widely applied, this process results in pollutant emissions and requires cleaning mechanisms. The generated thermal energy can be used directly in heating or for electrical generation.
- Pyrolysis—consists of waste transformation through its decomposition of medium or high temperature (from 200°C to 1100°C) and the absence of oxygen, resulting in bio-oil (through condensation), coal, and noncondensable gases. The share of each of these products will depend on the temperature and speed of the reaction. Pyrolysis occurs in a reactor and the most common form of heat transfer is through induction, although microwaves may also be used. Waste pretreatment mainly involves drying and crushing.

- Gasification—process that converts the waste into a gas, known as synthesis gas, composed mainly of CO, H_2 and CO_2. The transformation takes place in a gasifier in an environment with controlled oxygen mixing. The literature identifies three types of gasifiers: fixed bed, fluidized bed, and entrained bed. Pretreatment for gasification involves drying and grinding the residue. The synthesis gas can be used for direct burning, can be converted into biofuel or used for electrical generation. In addition to this gas, the process results in two solid coproducts: tar and coke.
- Anaerobic digestion—applicable to biogenic and easily biodegradable waste, which is converted into biogas, composed mainly of methane and carbon dioxide. Anaerobic digestion occurs in fermentation tanks in the absence of oxygen and requires a pretreatment for crushing, homogenization, purification, and addition of nutrients. The methane resulting from the process can be used for electrical generation, for heating through combustion or as a convertible to biofuel. The by-product of anaerobic digestion can be used for soil fertilization.

Most of the publications focused on LCA for the energy use of waste include incineration in some way, which reflects the fact that this is the most applied energy conversion technology worldwide. Several studies point to the predominance of incineration, such as [8] who in a survey of 250 publications identified that 82% of them were focused on this technology. On the other hand, more advanced technologies, such as gasification and pyrolysis, have received increasing attention in more recent studies [6]. Even within the same type of technology, it is possible to identify the preference for a specific subtype. Among the studies that evaluate incineration technology, for example, a large part is based on mobile grid incinerators.

LCA study applications are linked to the definition of the objective and scope, which is one of the stages of the study, as indicated previously. Many studies use the LCA to compare between different technologies for the use of waste energy, or between the use of energy and other forms of waste management, while in some cases the LCA was applied to advance specific processes in the treatment units, in order to optimize energy use (Fig. 5.3).

5.4 Highlights in LCA studies for waste-to-energy technologies

5.4.1 Functional unit

The FU is the unit to which all other flows of matter and energy will be related. In LCA studies for WtE technologies, the FU will generally be a unit of energy (e.g., 1 MJ) generated at the end of the process or a unit of waste (i.e., 1 tonne) used as input. Thus if the energy unit is used, for example, all other calculated quantities will be related to that amount of produced energy.

The vast majority of studies adopt a certain amount of treated waste as an FU, which characterizes an input-based approach, while about 10% of the analyses establish

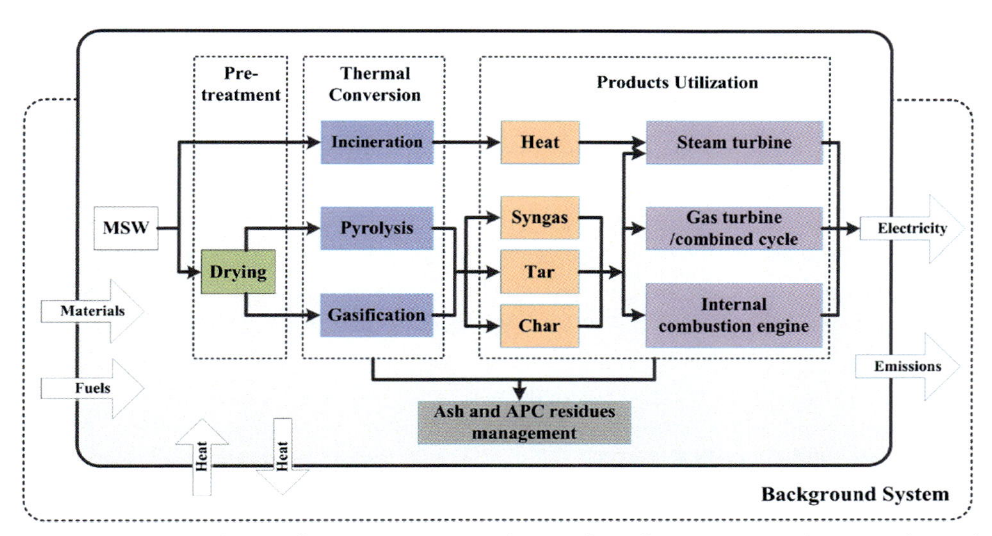

Figure 5.3 Example of a LCA for incineration, pyrolysis and gasification. *Source: (Dong et al., 2018).*

an amount of energy as an FU, in a product-based approach [6]. This predominance of the FU based on the entry of waste in the treatment unit reveals a perspective more focused on waste management [8]. Considering energy systems as a whole and not just WtE, a common practice is also the definition of the FU as the supply of a quantity of energy based on the demand of a country, a region or an institution [9].

5.4.2 Type of residue

The type of waste used in the study is important, as it will determine the quality of the generated fuel, as well as the efficiency and emissions resulting from the process. A wide range of waste types can be considered in the LCA of WtE technologies, from compound household waste to waste composed of a single type of material, which can be grouped as organic waste (biogenic) or as inorganic waste (anthropogenic).

Many LCA studies have been carried out based on "mixed municipal waste," that is, with different types of materials, which reflects the context of developed countries, where this type of waste has a high participation. On the other hand, developing countries present a significant fraction of biomass residues from agricultural activities. Thus a limitation in comparing the results of the various studies according to different geographical scopes is observed.

A smaller number of studies are based on residues composed of only one specific material. It is essential that the study provides information on the used waste composition and its chemical characteristics (e.g., lower calorific value) so transparency and the possibility of checking and/or comparing the results are available.

5.4.3 Form of energy use

There are different ways in which an energy recovery system can export the energy recovered in the process. For example, thermal energy may be used in residential heating or in industries, electricity generated in a turbine or engine can be injected into the grid, or as a liquid, gaseous or solid fuel, which can be used in different ways [6].

Electricity has been the most prominent, followed by heat, mainly in incineration [6,8]. When gasification is applied technology, electricity is again the main product, considered in 83% of all studies, but a higher frequency of fuels (25%) than heat (20%) is noted. The synthesis gas resulting from the process is generally used in engines or turbines. It is also noteworthy that, in the case of anaerobic digestion and hydrothermal carbonization, many studies still consider certain coproducts, such as digesters (digestate) and charcoal (biochar) [6].

5.4.4 Energy recovery

Energy recovery in a waste energy conversion system means taking advantage of the energy flows that leave the system, which would be lost, especially heat generated by the equipment. This is one of the main technical aspects in LCA studies applied to WtE and is crucial for the results.

Several studies include some form of energy recovery, the most common in the form of electricity. Among these, about half of the studies report the existence of a net electrical efficiency in the system. On the other hand, few LCA studies have included energy recovery in the form of heat. However, in this type the result seems to be better, since almost 60% report a net heat recovery [8].

5.4.5 Sensitivity and uncertainty analyses

Sensitivity and uncertainty analyses are very important steps in several LCA studies. A sensitivity analysis is used to indicate how much the results of the study may vary when some basic variables within the model are changed. For example, what should be the change in the impacts observed when the humidity level of the waste is increased by 10%, or when the oxygen mixture in the gasifier is reduced by 5%. On the other hand, an uncertainty analysis is related to the possibility that the applied parameters or scenarios are not correct, and it is necessary to evaluate how much this may affect the results.

Although they are recommended by international guidelines for conducting LCA studies, few studies related to WtE technologies carry out sensitivity and uncertainty analyses. Consequently, this reduces the robustness of the results obtained in these studies and makes them questionable.

5.5 Main results found in the literature

Waste treatment has been the subject of a high number of LCA studies in recent decades. Although the comparison between the performances of the different WtE technologies obtained in different studies is not highly recommended, due to methodological differences, some general results can be extracted from the literature. In fact, the comparison of impacts between technologies should ideally be performed in the same study, thus ensuring the use of the same methodology and the same type of waste [6].

When comparing WtE technologies with other forms of disposal, studies in general confirm the hierarchy of waste management, according to which recycling is preferable to energy recovery, which, in turn, is preferable to disposal in landfills. Between the first two options, recycling and energy recovery, some studies indicate that this may be better for waste paper and plastic, depending on the applied parameters. Similarly, between energy recovery and landfill disposal, a few studies pointed out that the latter is more advantageous for some types of materials and under specific conditions. These studies in general were based on modern landfills that provide significant reductions in CO_2 emissions and other environmental impacts [8].

Regarding the comparison between WtE technologies, many studies that compare gasification and pyrolysis, which are more advanced ways of using waste, by incineration and direct combustion in general, concluded that the first two forms of treatment are preferable [8]. These latter forms of energy conversion, especially incineration, have in fact been the target of criticism for their socio-environmental impacts, with emphasis on air pollution in the immediate surroundings.

Gasification can result in better environmental performance in some impact categories even when compared to anaerobic digestion, which is usually one of the best treatment options. The advantage of gasification can be verified, for example, in terms of global warming, photochemical oxidation and particulate materials. Compared with direct combustion, gasification performs better in several impact categories [10].

An issue that has been raised in studies published in the literature concerns the inclusion of capital goods at the borders of the system. In most cases, these assets have been excluded from the analysis or were not specified, and one of the justifications is that their contribution to the impacts is insignificant when compared to the operation phase [6,8]. However, more recent studies have questioned this statement, especially when observing specific impact categories, such as depletion of natural resources, eutrophication and toxicity [8]. In processes in which emissions during operation are lower, such as gasification, the infrastructure, including production, transportation and maintenance of machinery and equipment, tends to contribute more to impacts [10].

The production scale is also an important factor when deciding on the best WtE technology to be used and has a significant influence on costs [10]. For low levels of production, economies of scale and increased efficiency of the process as a whole are observed, but after a certain point, the effect is reversed mainly due to the need for biomass collection in an increasingly larger radius [11].

Although it was possible to extract some results from the literature, in general the various studies present very different panoramas regarding the sustainability of WtE technologies. This difference in results is related to the different approaches, technological standards and waste compositions adopted in the analyses. While the problems regarding the last two aspects cannot be solved, as they reflect the different geographical contexts, the issue of approaches can be overcome with greater rigor by the use of ISO standards, increasing the possibility of comparison between studies [6].

5.6 Conclusion

As noted initially, technologies for the use of waste energy (WtE technologies) display significant advances due to the various benefits that they can provide. The main one is probably the generation of fuels or electricity based on residues to replace fossil fuels, thus contributing to the reduction of GHG emissions and the mitigation of climate change. In addition to the climate issue, the use of energy generation contributes to a more adequate waste management, in addition to being an alternative to offer electricity in places exhibiting difficult access to this form of energy. However, despite all these benefits, these technologies also present the potential to cause environmental impacts and therefore the pros and cons of their use must be assessed.

The LCA is a methodology used to assess the environmental impacts of products and processes, with the main characteristic of being comprehensive, ideally including all stages from the extraction of raw materials to their final destination. Developed from the 1960s for some specific cases, LCA has undergone several methodological advances since then and has been applied in several areas, including waste management and WtE technologies.

The analysis of the literature on LCA applied to WtE technologies indicates that some aspects display greater importance, for which more attention is needed in this type of assessments, such as the definition of the FU, the type of assessed waste, the way of using the energy generated in the process, and the possibility of recovering the energy that is lost outside the system. In addition, sensitivity analyses for the parameters used in the evaluation are paramount.

The results found in the literature, in general, confirm the order presented in the waste management hierarchy, according to which recycling is more advantageous than energy use, which in turn is more advantageous than final disposal in landfills or dumps. Thus the LCA of WtE technologies demonstrayes that, although they can also cause impacts, the benefits are greater, making them viable from an environmental point of view. In addition, more advanced technologies, such as gasification and pyrolysis, display better environmental performance when compared to incineration, which is a much more widespread form of recovery. Finally, the use of the LCA is useful for promoting improvements in these technologies since it allows for the identification of the stages responsible for the greatest impacts throughout the life cycle.

References

[1] Ardolino F, Lodato C, Astrup TF, Arena U. Energy recovery from plastic and biomass waste by means of fluidized bed gasification: a life cycle inventory model. Energy 2018;165:299–314. Available from: https://doi.org/10.1016/j.energy.2018.09.158.

[2] Guinée JB, Heijungs R, Huppes G, Zamagni A, Masoni P, Buonamici R, et al. Life cycle assessment: past, present and future. Environ Sci Technol 2011;45:90–6. Available from: https://doi.org/10.1021/es101316v.

[3] Owsianiak M, Bjørn A, Laurent A, Molin C, Ryberg MW. LCA applications. In: Hauschild MZ, Rosenbaum RK, Olsen SI, editors. Life cycle assessment: theory and practice. Cham: Springer International Publishing; 2018, p. 31–41. Available from: https://doi.org/10.1007/978-3-319-56475-3_4.

[4] Owsianiak M, Bjørn A, Laurent A, Molin C, Ryberg MW. LCA applications. life cycle assessment: theory and practice. Springer International Publishing; 2017, p. 31–41. Available from: https://doi.org/10.1021/es101316v. Available from: https://doi.org/10.1007/978-3-319-56475-3_4.

[5] Castelo Branco DA. Avaliação do Real Potencial de Redução das Emissões de CO2 Equivalente com Uso da Captura em uma UTE a Carvão. Tese D Sc Programa de Planej Energético—COPPE/UFRJ; 2012.

[6] Mayer F, Bhandari R, Gäth S. Critical review on life cycle assessment of conventional and innovative waste-to-energy technologies. Sci Total Environ 2019;672:708–21. Available from: https://doi.org/10.1016/j.scitotenv.2019.03.449.

[7] Laurent A, Espinosa N, Hauschild MZ. LCA of energy systems. In: Hauschild MZ, Rosenbaum RK, Olsen SI, editors. Life cycle assessment: theory and practice. Cham: Springer International Publishing; 2018, p. 633–68. Available from: https://doi.org/10.1007/978-3-319-56475-3_26.

[8] Astrup TF, Tonini D, Turconi R, Boldrin A. Life cycle assessment of thermal waste-to-energy technologies: review and recommendations. Waste Manage 2015;37:104–15. Available from: https://doi.org/10.1016/j.wasman.2014.06.011.

[9] Chen H, Yang Y, Yang Y, Jiang W, Zhou J. A bibliometric investigation of life cycle assessment research in the web of science databases. Int J Life Cycle Assess 2014;19:1674–85. Available from: https://doi.org/10.1007/s11367-014-0777-3.

[10] Aberilla JM, Gallego-Schmid A, Azapagic A. Environmental sustainability of small-scale biomass power technologies for agricultural communities in developing countries. Renew Energy 2019;141:493–506. Available from: https://doi.org/10.1016/j.renene.2019.04.036.

[11] Yang K, Zhu N, Yuan T. Analysis of optimum scale of biomass gasification combined cooling heating and power (CCHP) system based on life cycle assessment (LCA). Procedia Eng 2017;205:145–52. Available from: https://doi.org/10.1016/j.proeng.2017.09.946.

CHAPTER 6

Waste disposal in selected favelas (slums) of Rio de Janeiro

6.1 Historical background

Historically, the Providencia hill (Morro da Providência) is considered the first favela in Rio de Janeiro.

In the second half of the 19th century, with the expansion of Rio de Janeiro, then headquarters of the federal government, not only in urban but also in peripheral areas, an increasing number of workers arrived. This is associated with the political independence achieved in 1822, the coffee reign, and the abolition of slavery that aided a phase of economic expansion.

The misaligned population growth has led to the growth of precarious housing and worsening manifestations of poverty. The collective dwellings known as tenements were multiplied.

In 1897 around 20,000 soldiers who had returned to Rio de Janeiro after the Canudos War in the eastern province of Bahia began living in the already inhabited Morro da Providência. During the conflict, the ruling troop had focused on the Bahian hinterland in the region near a hill called Favella, the name of hardy vegetation that caused irritation when in contact with human skin and common in the region. Because it sheltered people who had fought in that conflict, Morro da Providência was then nicknamed "Morro da Favela." The name became popular and, from the 1920s, the hills covered by shacks and huts were called favelas [1].

However, as the recently proclaimed capital of the Republic of Brazil, Rio de Janeiro needed to undergo renovations to become a more European and modern city by the standards of the time. It was then that Mayor Francisco Pereira Passos began to make extensive urban renovations in the city center, which included the expansion and opening of new roads, such as Avenida Central. During the renovations, several tenements were demolished, and their residents forced to look for other ways to live in the increasingly valued center, including the nearby hills, which forced a strong expansion of the favelas in the period. However, residents of these settlements would only be recognized by society and the government from the 1920s [1].

In addition, in the Estado Novo era, under the government of Getúlio Vargas, and Carlos Lacerda in Rio de Janeiro, until the Military Regime in the 1960s, several slums' removal and disposal programs evicted and displaced thousands of people and

A Thermo-Economic Approach to Energy From Waste
DOI: https://doi.org/10.1016/B978-0-12-824357-2.00003-6

destroyed several shacks, showing no success in solving the problem. Between 1962 and 1974 almost 80 slums were involved in these programs, resulting in 26,193 destroyed shacks and 139,218 inhabitants removed. During the military dictatorship, some leaders of favela communities were tortured and killed [2].

In the early 1980s with population swelling, the absence of the state and the consequent lack of public policies, favelas became the main centers of drug trafficking in Rio de Janeiro, which made these areas even more violent [3].

It was in 1983 under Brizola's government, when these settlements were already consolidated and their populations already huge, that the municipal government began to look for ways to urbanize the city's slums, rather than simply toppling them. During this period, programmes, such as Favela-Bairro, came to bring some kind of infrastructure to these areas, such as running water, basic sanitation, garbage collection, and public lighting.

In 2008 Rio de Janeiro's State Public Security Bureau started to implement the project of the Pacifying Police Unit (UPP), which consists of establishing community police units in slums dominated by drug trafficking, taking back control of the territory for the State. Currently, the project, which has benefited about 280,000 people in the city, still exists but has lost strength. The first UPP was installed at Favela of Santa Marta on November 20, 2008. Later, other units were implemented at Cidade de Deus, Batam, Pavão-Pavãozinho, and Morro dos Macacos, among other favelas [4].

Despite several criticisms because the first communities to receive UPPs were located near the richest part of the city, the program has been well evaluated by experts as a way to reduce crime in the wealthier neighborhoods.

Faced with this perspective, the City Hall initiated the Morar Carioca Program, intended to guide the urbanization and integration of favelas. It was introduced in 2010 and with the aim of better results the City Hall undertook a thorough review of how it perceives, registers, and operates in its slum areas [4].

6.1.1 Some numbers about subnormal clusters

The first data on subnormal clusters—corresponding to slums or favelas in the case of Rio de Janeiro—were released in the 2010 census only in December 2011 by the IBGE (Brazilian Institute of Geography and Statistics).

In the 2010 census, subnormal clusters (AGSN) accounted for 6% (11.4 million) of the Brazilian population and distributed in only 323 municipalities (6% of the total number). Almost half of this contingent was in the Southeast region, especially in the states of São Paulo (2.7 million) and Rio de Janeiro (2.0 million). Pará, with 1.2 million residents in AGSN, had the third largest number, although its population was only the ninth largest in the country. Thus favelas are an urban and metropolitan phenomenon: 88% of households in AGSN are concentrated in 20 of the 36 metropolitan regions and 45% in 15 municipalities with more than one million inhabitants. Rio is the national leader with about 1.4 million AGS residents,

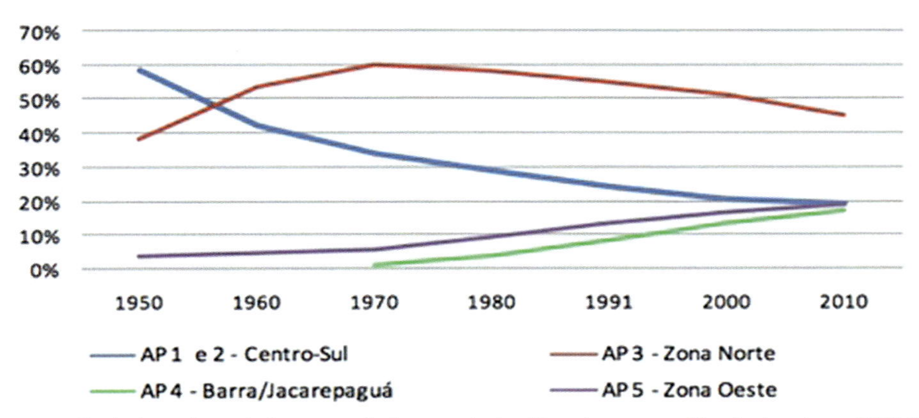

Figure 6.1 Evolution of total slum population yearly by Planning area - Rio de Janeiro - 1950-2010. Note: To alleviate inaccuracies, data from AP1, of low expression, were aggregated to those from AP2. *Own elaboration from IBGE Censo Demográfico 2010—Aglomerados Subnormais—primeiros resultados, Rio de Janeiro, RJ: IBGE; 2011 [5].*

immediately followed by São Paulo, with 1.3 million. Proportionally, however, Rio has 22% of its inhabitants in this condition.

According to [5], population living in slums in 2010 represented 23% of the total population of Rio de Janeiro with 1443 inhabitants. The proportions varied significantly between the regions of the city, with a predominance of Central Planning Area 1 (AP1)—which is located in neighborhoods, such as Catumbi, Rio Comprido, São Cristóvão, Santa Teresa, and Mangueira. As shown in Fig. 6.1, although in absolute terms the slum population of AP1 is the smallest in the city (about 103,000 inhabitants), its proportion of the total population in the area was the largest—35%.

It is worth mentioning that later, with about 26/27%, appear the North Zone and Barra/Jacarepaguá, with contingents much larger than in AP1: 654,000 and 236,000 residents, respectively, lived in slums. The smallest and almost identical proportions of slum dwellers were located in the West (16%) and South (17%) Zones. In absolute terms, however, the 274,000 slums in the West Zone represented about 100,000 more than in the South Zone (wealthiest zone).

6.1.2 The favela of Catumbi

Catumbi is a term derived from the Tupi language meaning blue leaf. Its origin is a village on the banks of the Catumbi River, central region of the city of Rio de Janeiro. Bordering the neighborhoods of Estácio, Cidade Nova, Centro, Santa Teresa, and Rio Comprido is one of the oldest favelas in the city. In the early days, it was a damp, shady valley through which a river born on the heights of Santa Teresa Hill ran, a river used for the irrigation of sugarcane crops [4].

In the 20th century, with the expansion of the urban fabric in other directions, the neighborhood went into decay. In 1964 the construction of the Santa Bárbara Tunnel contributed to this process, transforming the neighborhood into a passage corridor, a situation aggravated in the following decades by the process of swelling of the low-income neighboring communities. At the time, some factories settled in the neighborhood, such as the Old Brahma Factory on Marques de Sapucaí Street, where nowadays the current popular Carnival presentation takes place [4].

According to 2010 Census, the population of this neighborhood is at 12,556 inhabitants. In Catumbi, there are more women than men, with a population composed of 53.51% women and 46.49% men [5].

Fig. 6.2 shows age groups, in categories from 0 to 4 years, 0 to 14 years, 15 to 64 years, and 65 years and + . In sight, there are younger people than elderly. Population is made up of 20.8% young people and up to 10.1% elderly people.

Fig. 6.3 shows that regarding the level of education, 52.8% of respondents from favela of Catumbi have incomplete elementary school. Higher education, unfortunately, only reaches the smallest 2.8% of such households.

Fig. 6.4 shows that in the favela of Catumbi, 45.8% of the households consist of 4–5 residents.

Fig. 6.5 shows that 48.6% of respondents from the favela of Catumbi receive a monthly wage from 678.00 (161 U$D) to 1356.00 (322 U$D) Reais, that is to say, values lower than two minimum wages at the time in Brazil.

In Fig. 6.6, we can see that 40.3% of the family monthly income range is up to 2034 Reais (484 U$D).

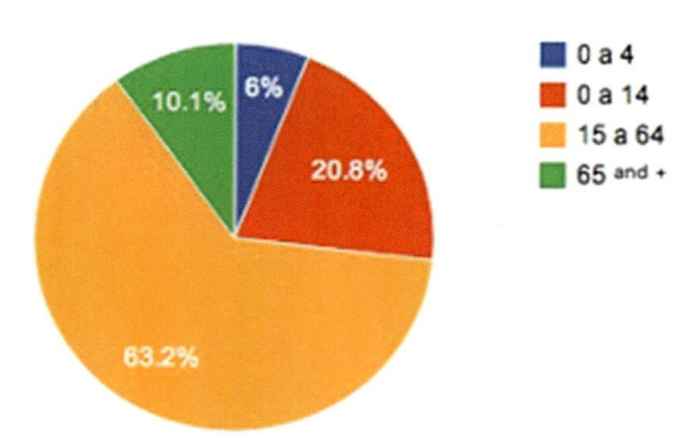

Figure 6.2 Age range of Catumbi - Rio de Janeiro's population. *Own elaboration from IBGE Censo Demográfico 2010—Aglomerados Subnormais—primeiros resultados, Rio de Janeiro, RJ: IBGE; 2011 [5].*

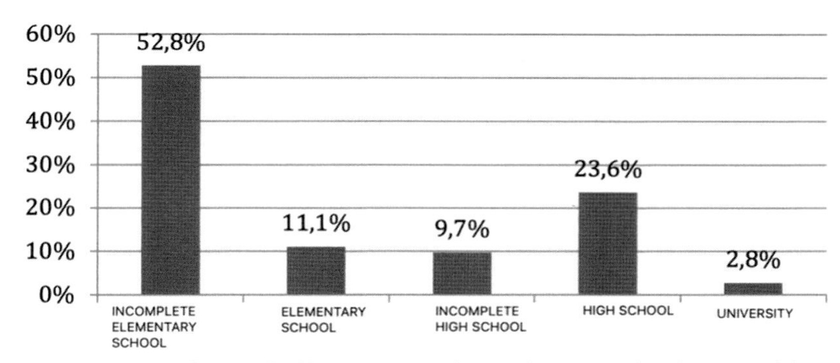

Figure 6.3 Distribution of households by respondent education's level. *Own elaboration from CENTROCLIMA/COPPE/UFF/SMAC. Relatório Final do Projeto Diagnóstico Preliminar de Resíduos Sólidos da Cidade do Rio de Janeiro; 2015 [6].*

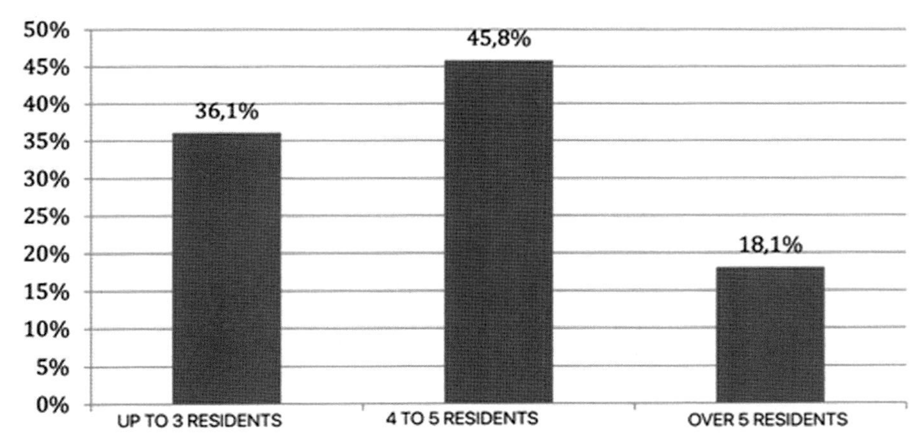

Figure 6.4 Number of residents per household. *Own elaboration from CENTROCLIMA/COPPE/UFF/SMAC. Relatório Final do Projeto Diagnóstico Preliminar de Resíduos Sólidos da Cidade do Rio de Janeiro; 2015 [6].*

6.2 Survey and study of solid waste in 37 slums and in Catumbi

Presenting more specifically the steps of the methodology adopted, initially an exploratory study and in-depth interviews were conducted with the following interlocutors: a representative and a resident of the favela. The information generated by the participants—collected through a script—was recorded and later analyzed according to the manner in which the script was applied.

In-person interviews, a fundamental part of quantitative research, have a margin of error of 3% in confidence interval calculations of 95% in proportion estimates—considering the worst case.

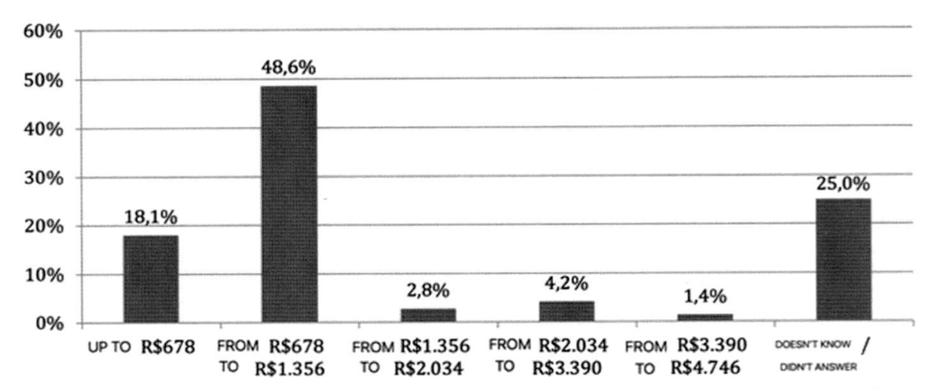

Figure 6.5 Distribution of households by monthly income range of respondents. *Own elaboration from CENTROCLIMA/COPPE/UFF/SMAC. Relatório Final do Projeto Diagnóstico Preliminar de Resíduos Sólidos da Cidade do Rio de Janeiro; 2015 [6].*

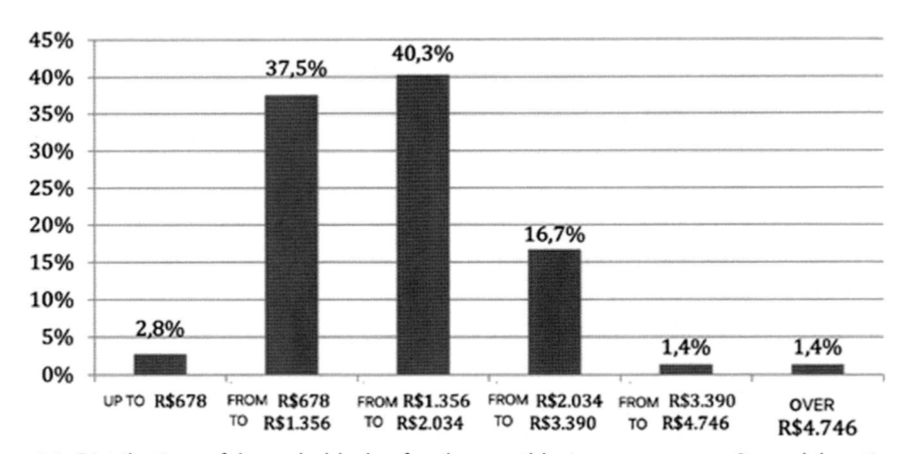

Figure 6.6 Distribution of households by family monthly income range. *Own elaboration from CENTROCLIMA/COPPE/UFF/SMAC. Relatório Final do Projeto Diagnóstico Preliminar de Resíduos Sólidos da Cidade do Rio de Janeiro; 2015 [6].*

As potential interviewers in the community, were chosen people who turn to residents to solve several types of problems and thus have a chance to have a broader and contextualized view of their problems, such as:

- Health post doctors,
- Community schools directors,
- Representatives of residents associations,
- Religious leaders,
- NGOs working with recycling, and
- NGOs working with culinary exploitation of leftovers.

To these potential respondents were also added some street cleaners.

After the interview and questionnaire steps, the starting point was the accounting of solid waste generated through the disposal of products purchased from monthly purchases of surveyed households. Thus the volume of waste generated was verified by the quantity of products purchased and their weight, also considering the type of material in which they fit—plastic, paper, glass, metal, Tetra Pak, and organic.

To this end, the weight of the packaging of organic products commonly purchased by households through empirical research and via specialized internet sites was lifted.

Given the great difficulty of lifting all the weights of the products involved in the search through specialized websites, the empirical method with a CAMRY brand digital scale model EK6550-06, steel material, and CE/EU certified (Fig. 6.7) was used.

For organic products, it should be noted that, for disposal, only those parts that are not consumed in general, such as bark, stones, and peduncles, were considered, depending on the type of product. To this end, a technical visit was made to a free market, where the discard weighing—peels, bark, seeds, and stalks, among others—of a unit consumed of fruits and vegetables present in household purchases was applied. In addition to contacts made via telephone/email and technical visit, it was necessary to consult other sources to obtain the weight of the various parts of the products. In the organic category websites of the Brazilian Micro and Small Business Support Service [7] and online shopping sites, Feira Feita and Hortifruti [8,9] and two papers by [10,11] were consulted. In the packaging category, Fibrasa and Plastikero [8,12] were consulted to obtain the weight of their products. On the other hand, there are also products that fit into both categories. For example, for toilet paper and paper towels are considered: the disposal of the packaging (plastic) and the product itself (paper, but categorized as organic, given the nature of its use), which also applies to intimate tampons.

Figure 6.7 Digital scale, brand CAMRY, model EK6550-06.

The methodology for obtaining the total waste generated (in kg) is:

$$\text{For recyclables: } \mathrm{Pe}(\mathrm{Pe} \times E) \times D$$

$$\text{For organic: } \sum (\mathrm{Pc} \times A) \times D$$

where Pe is the *package*-specific weight in g, E is the packaging in units, Pc is the measured weight for food residues in g, A is the food in units, and D is the total number of households surveyed.

$$\text{Total waste generated(TR)} = \left(\sum (\mathrm{Pe} \times E) \times D\right) + \left(\sum (\mathrm{Pc} \times A) \times D\right)$$

Then, the volume of waste disposed of irregularly was accounted for, based on a specific question from the questionnaire: number 18 (where do you dispose of your household waste?). To obtain the total volume of irregularly disposed waste (TRI) per community, it was considered:

$$V_{\mathrm{RI}} = \frac{D - (R_1 + R_2)}{D} \times 100$$

$$T_{\mathrm{RI}} = \frac{T_R \times V_{\mathrm{RI}}}{100}$$

where V_{RI} is the percentage of households that answered yes to incorrect waste disposal, R_1 is the total households disposing of their waste in the waste basket, R_2 is the total households disposing of their waste in the dumpster, and IRT is the total volume of irregularly disposed waste.

Table 6.1 presents the total monthly amount of waste generated in the 37 surveyed favelas, as well as the per capita and per household values per day. We can see that Catumbi is the third largest generator in the monthly total, with 3490.56 kg; Rio das Pedras is the second (4034.60 kg), and then, Rocinha is leading with 5568.42 kg of discarded materials per month. Table 6.1 also shows that Fazenda Botafogo leads the production of waste per household per month, although Parque Tiradentes is the leader in per capita generation of total waste per month.

Table 6.2 shows that, given the socioeconomic profile of the vast majority of slum dwellings, organic waste leads to the total amount of waste, corresponding to 80% of the total generated. Glass and plastic are the most discarded material categories after organic waste.

As shown in Table 6.3, the volume generated by total material and per household for *Catumbi specifically* also shows the predominance of organic material, followed by plastic.

Table 6.1 Total amount of waste generated in the slums surveyed: total per month, per household, and per capita (kg).

	Weight (kg)					
Favelas	Total	Per household	Per capita	Total per day	Per household per day	Per capita per day
Rocinha	5568.42	54.06	17.79	185.61	1.80	0.59
Rio das Pedras	4034.60	49.20	14.01	134.49	1.64	0.47
Jacarezinho	2085.39	45.33	14.48	69.51	1.51	0.48
Parque União	633.14	19.79	8.22	21.10	0.66	0.27
Unidos de Santa Tereza	151.65	15.17	6.59	5.06	0.51	0.22
Morro dos Telégrafos	1072.03	48.73	11.28	35.73	1.62	0.38
Vila do Vintém	1328.97	57.78	13.99	44.30	1.93	0.47
Morro Santa Marta	763.56	38.18	11.07	25.45	1.27	0.37
Morro do Alemão	859.98	45.26	13.65	28.67	1.51	0.46
Nova Cidade	1108.48	48.19	14.03	36.95	1.61	0.47
Parque Maré	2538.96	52.89	17.04	84.63	1.76	0.57
Canal do Cortado	480.17	20.01	6.40	16.01	0.67	0.21
Estrada da Pedra IV	334.09	13.92	6.68	11.14	0.46	0.22
Tijuquinha	934.46	38.94	13.16	31.15	1.30	0.44
Vila dos Crentes	392.11	16.34	4.84	13.07	0.54	0.16
Rio Morto	431.79	17.99	4.32	14.39	0.60	0.14
Jardim Gramado	1213.29	50.55	12.26	40.44	1.69	0.41
Vila Arará	871.00	36.29	11.77	29.03	1.21	0.39
Bosque Mont Serrat	626.13	26.09	7.28	20.87	0.87	0.24
Comunidade Agrícola de Higienópolis	1164.57	48.52	15.53	38.82	1.62	0.52
Alto Bela Vista	1220.33	50.85	14.70	40.68	1.69	0.49
Parque São Sebastião	1093.65	45.57	13.34	36.45	1.52	0.44
Fazenda Botafogo/ Margem da Linha	1516.97	63.21	15.48	50.57	2.11	0.52
Pantanal	1453.89	60.58	14.39	48.46	2.02	0.48
Catumbi	3490.56	48.48	11.67	116.35	1.62	0.39
Morro do Cariri	1126.62	46.94	15.87	37.55	1.56	0.53
Parque Alvorada	483.80	20.16	6.12	16.13	0.67	0.20
Santa Terezinha	591.14	24.63	7.48	19.70	0.82	0.25
Borel	1326.84	55.29	15.25	44.23	1.84	0.51
Parque João Paulo II	1048.68	43.70	13.27	34.96	1.46	0.44
Vidigal	1022.97	42.62	12.03	34.10	1.42	0.40
Parque Vila Isabel	1256.06	52.34	14.61	41.87	1.74	0.49
Pavão-Pavãozinho	1468.47	61.19	13.35	48.95	2.04	0.44
Muzema	487.02	20.29	5.87	16.23	0.68	0.20
Parque Tiradentes	1333.19	55.55	18.02	44.44	1.85	0.60
Total	48,207.54	43.83	12.74	1606.92	1.46	0.42

Own elaboration from CENTROCLIMA/COPPE/UFF/SMAC. Relatório Final do Projeto Diagnóstico Preliminar de Resíduos Sólidos da Cidade do Rio de Janeiro; 2015 [6].

Table 6.2 Total amount of slum waste by material per month.

	Weight (kg)		
Material type	**Total**	**Per household**	**Per capita**
Plastics	2875.61	2.61	0.76
Tetra Pak	1005.71	0.91	0.27
Paper	1808.80	1.64	0.48
Glass	2728.47	2.48	0.72
Organics	38,855.73	35.32	10.27
Metal	933.23	0.85	0.25
Total	48,207.54	43.83	12.74

Own elaboration from CENTROCLIMA/COPPE/UFF/SMAC. Relatório Final do Projeto Diagnóstico Preliminar de Resíduos Sólidos da Cidade do Rio de Janeiro; 2015 [6].

Table 6.3 Volume of total and household waste generated by material type per month—Catumbi.

	Weight (kg)	
Material type	**Total**	**Per household**
Plastics	227.95	3.17
Tetra Pak	95.13	1.32
Paper	91.49	1.27
Glass	131.77	1.83
Organics	2865.06	39.79
Metal	79.17	1.10
Total	3490.56	48.48

Own elaboration from CENTROCLIMA/COPPE/UFF/SMAC. Relatório Final do Projeto Diagnóstico Preliminar de Resíduos Sólidos da Cidade do Rio de Janeiro; 2015 [6].

The total waste generated in the Catumbi sample was approximately 3491 kg per month, with a value per household of about 48.5 kg per month and a daily value per household of 1.6 kg. Considering the average of four residents per household, per capita generation obtained corresponds to 0.4 kg per day. This number is not a significant one in relation to general per capita waste generation according to the World Bank—which is 1.2 kg per person per day. It should be noted that the consumption profile and income pattern of Catumbi considerably reduce the amount of packaging and increase the organic generation.

As can be seen from Table 6.4, when compared to organic waste, Catumbi households compared to Rocinha households—the largest generator in this category for total households (11% of the total generated monthly)—represent 63.20% of the total of Rocinha, a very significant percentage in this sample universe. It is also observed

Table 6.4 Organic material per favela: total per month, per household, and per capita.

Favelas	Weight (kg)		
	Total	Per household	Per capita
Rocinha	4532.98	44.01	14.48
Rio das Pedras	3265.56	39.82	11.34
Jacarezinho	1828.03	39.74	12.69
Parque União	439.84	13.75	5.71
Unidos de Santa Tereza	104.82	10.48	4.56
Morro dos Telégrafos	938.33	42.65	9.88
Vila do Vintém	1154.36	50.19	12.15
Morro Santa Marta	652.32	32.62	9.45
Morro do Alemão	765.04	40.27	12.14
Nova Cidade	964.67	41.94	12.21
Parque Maré	2248.69	46.85	15.09
Canal do Cortado	341.38	14.22	4.55
Estrada da Pedra IV	269.11	11.21	5.38
Tijuquinha	469.15	19.55	6.61
Vila dos Crentes	314.19	13.09	3.88
Rio Morto	320.07	13.34	3.20
Jardim Gramacho	1066.32	44.43	10.77
Vila Arará	755.15	31.46	10.20
Bosque Mont Serrat	466.44	19.44	5.42
Comunidade Agrícola de Higienópolis	890.09	37.09	11.87
Alto Bela Vista	970.97	40.46	11.70
Parque São Sebastião	954.83	39.78	11.64
Fazenda Botafogo/Margem da Linha	1162.31	48.43	11.86
Pantanal	1130.00	47.08	11.19
Catumbi	2865.06	39.79	9.58
Morro União	1150.47	47.94	11.74
Chico Mendes (Morro do Chapadão)	993.88	41.41	11.83
Morro do Cariri	802.53	33.44	11.30
Parque Alvorada	332.06	13.84	4.20
Santa Terezinha	456.23	19.01	5.78
Borel	1014.48	42.27	11.66
Parque João Paulo II	802.99	33.46	10.16
Vidigal	904.25	37.68	10.64
Parque Vila Isabel	988.80	41.20	11.50
Pavão-Pavãozinho	1154.50	48.10	10.50
Muzema	335.55	13.98	4.04
Parque Tiradentes	1.050.30	43.76	14.19
Total	38,855.73	35.32	10.27

Own elaboration from CENTROCLIMA/COPPE/UFF/SMAC. Relatório Final do Projeto Diagnóstico Preliminar de Resíduos Sólidos da Cidade do Rio de Janeiro; 2015 [6].

that Parque Maré has the highest volume of per capita generation (15 kg/month) and Vila do Vintém generates more waste per household (50.19 kg/month).

Regarding plastics (Table 6.5), Catumbi represents, in the total generation of all slums, a percentage of 7.9. From an absolute point of view, plastic waste is more generated in Rocinha households, followed by Rio das Pedras. Although the lead in plastic waste generation per capita remains at Rocinha, Fazenda Botafogo is the favela that most discards plastics per household.

For Tetra Pak waste (Table 6.6), Catumbi leads the total generation with a value of 95.13 kg per month. Borel is the slum that produces most waste per capita and per household in this regard.

Table 6.7 presents the total generated of 1808.80 kg/month of paper, in which Catumbi represents, with its 91.49 kg/month, a percentage of 5 of this total. Tijuquinha leads to the total generation per month, per household, and per capita.

Glass (Table 6.8) is more discarded in Rocinha, followed by Rio das Pedras. Household generation of glass waste is higher at Fazenda Botafogo, although per capita generation is led by Morro do Cariri. Catumbi produces the equivalent of 33% of the glass waste dumped in the same period by Rocinha—leader in total generation in this segment.

Table 6.9 shows that metal waste is more produced in Rocinha (111.64 kg/month), followed by Catumbi with generation at 79.17 kg per month. Per household, this waste is most discarded at Fazenda Botafogo and the largest generation per capita is in Rocinha, followed by Alto Bela vista and Morro do Chapadão.

As shown in Table 6.10, the volume generated by total material and by household shows that there was a predominance of organic material, followed by plastic.

Fig. 6.8 shows that the disposal of garbage from Catumbi occurs in 56.9% in dumpsters, while 30.6% of households leave their garbage on the floor in front of their homes.

Last but not least, it was found from the field study that organic, recyclable, and special waste (construction debris, used furniture, and household appliances) are mistakenly discarded, with minor exceptions for cans and pet bottles, whose most lucrative appeal attracts some collectors and artisans, although detached from any project that encourages and organizes them.

This is mainly due to the ingrained habit of disposing of trash at traditional points prior to bucket placement. It is almost a mechanical conditioning, which needs not only constant but also creative educational actions to be deprogrammed.

Overall, the daily garbage collection was well evaluated. However, in the opinion of residents, broader, integrated, and multisectoral educational actions are lacking.

There are other factors to be examined, such as the lack of buckets and swamps in some favelas, especially in large ones, such as Rocinha and Rio das Pedras. This would aggravate the picture, as it would increase the unwillingness to walk to the correct point.

Table 6.5 Plastic waste by slum: total per month, per household, and per capita.

Favelas	Weight (kg)		
	Total	Per household	Per capita
Rocinha	318.48	3.09	1.02
Rio das Pedras	257.37	3.14	0.89
Jacarezinho	100.09	2.18	0.70
Parque União	65.41	2.04	0.85
Unidos de Santa Tereza	13.34	1.33	0.58
Morro dos Telégrafos	63.43	2.88	0.67
Vila do Vintém	64.68	2.81	0.68
Morro Santa Marta	40.71	2.04	0.59
Morro do Alemão	45.93	2.42	0.73
Nova Cidade	54.25	2.36	0.69
Parque Maré	115.38	2.40	0.77
Canal do Cortado	38.47	1.60	0.51
Estrada da Pedra IV	28.29	1.18	0.57
Tijuquinha	55.55	2.31	0.78
Vila dos Crentes	31.07	1.29	0.38
Rio Morto	34.54	1.44	0.35
Jardim Gramacho	60.01	2.50	0.61
Vila Arará	50.73	2.11	0.69
Bosque Mont Serrat	46.51	1.94	0.54
Comunidade Agrícola de Higienópolis	64.38	2.68	0.86
Alto Bela Vista	70.17	2.92	0.85
Parque São Sebastião	51.50	2.15	0.63
Fazenda Botafogo/Margem da Linha	97.65	4.07	1.00
Pantanal	91.99	3.83	0.91
Catumbi	227.95	3.17	0.76
Morro União	80.10	3.34	0.82
Chico Mendes (Morro do Chapadão)	82.93	3.46	0.99
Morro do Cariri	64.12	2.67	0.90
Parque Alvorada	43.64	1.82	0.55
Santa Terezinha	44.78	1.87	0.57
Borel	80.46	3.35	0.92
Parque João Paulo II	61.71	2.57	0.78
Vidigal	54.34	2.26	0.64
Parque Vila Isabel	68.42	2.85	0.80
Pavão-Pavãozinho	84.32	3.51	0.77
Muzema	51.21	2.13	0.62
Parque Tiradentes	71.72	2.99	0.97
Total	2875.61	2.61	0.76

Source: Own elaboration from CENTROCLIMA/COPPE/UFF/SMAC. Relatório Final do Projeto Diagnóstico Preliminar de Resíduos Sólidos da Cidade do Rio de Janeiro; 2015 [6].

Table 6.6 Tetra Pak waste by favela: total per month, per household, and per capita.

	Weight (kg)		
Favelas	**Total**	**Per household**	**Per capita**
Rocinha	83.53	0.81	0.27
Rio das Pedras	75.85	0.92	0.26
Jacarezinho	45.95	1.00	0.32
Parque União	14.38	0.20	0.08
Unidos de Santa Tereza	2.54	0.25	0.11
Morro dos Telégrafos	18.79	0.85	0.20
Vila do Vintém	28.37	1.23	0.30
Morro Santa Marta	19.09	0.95	0.28
Morro do Alemão	12.47	0.66	0.20
Nova Cidade	21.48	0.93	0.27
Parque Maré	48.92	1.02	0.33
Canal do Cortado	20.91	0.87	0.28
Estrada da Pedra IV	17.87	0.74	0.36
Tijuquinha	27.89	1.16	0.39
Vila dos Crentes	14.16	0.59	0.17
Rio Morto	12.76	0.53	0.13
Jardim Gramacho	23.73	0.99	0.24
Vila Arará	18.22	0.76	0.25
Bosque Mont Serrat	21.49	0.90	0.25
Comunidade Agrícola de Higienópolis	30.05	1.25	0.40
Alto Bela Vista	17.48	0.73	0.21
Parque São Sebastião	25.05	1.04	0.31
Fazenda Botafogo/Margem da Linha	20.46	0.85	0.21
Pantanal	32.53	1.36	0.32
Catumbi	95.13	1.32	0.32
Morro União	12.28	0.51	0.13
Chico Mendes (Morro do Chapadão)	18.13	0.76	0.22
Morro do Cariri	28.15	1.17	0.40
Parque Alvorada	16.66	0.69	0.21
Santa Terezinha	13.81	0.58	0.17
Borel	38.20	1.59	0.44
Parque João Paulo II	26.59	1.11	0.34
Vidigal	17.65	0.74	0.21
Parque Vila Isabel	13.94	0.58	0.16
Pavão-Pavãozinho	17.70	0.74	0.16
Muzema	26.35	1.10	0.32
Parque Tiradentes	27.16	1.13	0.37
Total	1005.71	0.91	0.27

Source: Own elaboration from CENTROCLIMA/COPPE/UFF/SMAC. Relatório Final do Projeto Diagnóstico Preliminar de Resíduos Sólidos da Cidade do Rio de Janeiro; 2015 [6].

Table 6.7 Paper waste by slum: total per month, per household, and per capita.

Favelas	Weight (kg)		
	Total	Per household	Per capita
Rocinha	122.78	1.19	0.39
Rio das Pedras	101.55	1.24	0.35
Jacarezinho	40.42	0.88	0.28
Parque União	32.31	1.01	0.42
Unidos de Santa Tereza	14.59	1.46	0.63
Morro dos Telégrafos	18.89	0.86	0.20
Vila do Vintém	26.02	1.13	0.27
Morro Santa Marta	25.68	1.28	0.37
Morro do Alemão	7.22	0.38	0.11
Nova Cidade	22.07	0.96	0.28
Parque Maré	41.87	0.87	0.28
Canal do Cortado	38.29	1.60	0.51
Estrada da Pedra IV	3.38	0.14	0.07
Tijuquinha	340.38	14.18	4.79
Vila dos Crentes	14.74	0.61	0.18
Rio Morto	27.96	1.16	0.28
Jardim Gramacho	21.16	0.88	0.21
Vila Arará	14.55	0.61	0.20
Bosque Mont Serrat	49.81	2.08	0.58
Comunidade Agrícola de Higienópolis	57.00	2.37	0.76
Alto Bela Vista	41.17	1.72	0.50
Parque São Sebastião	28.59	1.19	0.35
Fazenda Botafogo/Margem da Linha	44.19	1.84	0.45
Pantanal	47.97	2.00	0.47
Catumbi	91.49	1.27	0.31
Morro União	45.72	1.91	0.47
Chico Mendes (Morro do Chapadão)	32.30	1.35	0.38
Morro do Cariri	55.37	2.31	0.78
Parque Alvorada	35.69	1.49	0.45
Santa Terezinha	39.21	1.63	0.50
Borel	53.49	2.23	0.61
Parque João Paulo II	40.10	1.67	0.51
Vidigal	16.35	0.68	0.19
Parque Vila Isabel	57.23	2.38	0.67
Pavão-Pavãozinho	54.41	2.27	0.49
Muzema	53.39	2.22	0.64
Parque Tiradentes	51.50	2.15	0.70
Total	1808.80	1.64	0.48

Source: Own elaboration from CENTROCLIMA/COPPE/UFF/SMAC. Relatório Final do Projeto Diagnóstico Preliminar de Resíduos Sólidos da Cidade do Rio de Janeiro; 2015 [6].

Table 6.8 Glass waste by slum: total per month, per household, and per capita.

	Weight (kg)		
Favelas	**Total**	**Per household**	**Per capita**
Rocinha	399.01	3.87	1.27
Rio das Pedras	263.24	3.21	0.91
Jacarezinho	31.68	0.69	0.22
Parque União	59.23	1.85	0.77
Unidos de Santa Tereza	12.45	1.24	0.54
Morro dos Telégrafos	18.65	0.85	0.20
Vila do Vintém	29.95	1.30	0.32
Morro Santa Marta	10.52	0.53	0.15
Morro do Alemão	14.19	0.75	0.23
Nova Cidade	23.53	1.02	0.30
Parque Maré	48.92	1.02	0.33
Canal do Cortado	25.78	1.07	0.34
Estrada da Pedra IV	8.69	0.36	0.17
Tijuquinha	28.28	1.18	0.40
Vila dos Crentes	8.76	0.36	0.11
Rio Morto	27,93	1.16	0.28
Jardim Gramado	17.31	0.72	0.17
Vila Arará	20.50	0.85	0.28
Bosque Mont Serrat	28.22	1.18	0.33
Comunidade Agrícola de Higienópolis	99.26	4.14	1.32
Alto Bela Vista	91.36	3.81	1.10
Parque São Sebastião	18.95	0.79	0.23
Fazenda Botafogo/Margem da Linha	160.80	6.70	1.64
Pantanal	126.25	5.26	1.25
Catumbi	131.77	1.83	0.44
Morro União	92.27	3.84	0.94
Chico Mendes (Morro do Chapadão)	129.97	5.42	1.55
Morro do Cariri	153.13	6.38	2.16
Parque Alvorada	44.41	1.85	0.56
Santa Terezinha	26.24	1.09	0.33
Borel	114.92	4.79	1.32
Parque João Paulo II	96.92	4.04	1.23
Vidigal	12.82	0.53	0.15
Parque Vila Isabel	102.64	4.28	1.19
Pavão-Pavãozinho	130.55	5.44	1.19
Muzema	10.12	0.42	0.12
Parque Tiradentes	109.26	4.55	1.48
Total	2728.47	2.48	0.72

Source: Own elaboration from CENTROCLIMA/COPPE/UFF/SMAC. Relatório Final do Projeto Diagnóstico Preliminar de Resíduos Sólidos da Cidade do Rio de Janeiro; 2015 [6].

Some residents, even near the correct disposal site, put their trash bag across the street or even put it next to the hopper. This seems to contradict the idea that the existence or even the number of buckets is the most important factor in this equation.

Table 6.9 Metal waste by slum: total per month, per household, and per capita.

Favelas	Weight (kg)		
	Total	Per household	Per capita
Rocinha	111.64	1.08	0.36
Rio das Pedras	71.04	0.87	0.25
Jacarezinho	39.22	0.85	0.27
Parque União	21.98	0.69	0.29
Unidos de Santa Tereza	3.92	0.39	0.17
Morro dos Telégrafos	13.95	0.63	0.15
Vila do Vintém	25.59	1.11	0.27
Morro Santa Marta	15.25	0.76	0.22
Morro do Alemão	15.13	0.80	0.24
Nova Cidade	22.49	0.98	0.28
Parque Maré	35.18	0.73	0.24
Canal do Cortado	15.34	0.64	0.20
Estrada da Pedra IV	6.76	0.28	0.14
Tijuquinha	13.21	0.55	0.19
Vila dos Crentes	9.20	0.38	0.11
Rio Morto	8.53	0.36	0.09
Jardim Gramado	24.76	1.03	0.25
Vila Arará	11.85	0.49	0.16
Bosque Mont Serrat	13.66	0.57	0.16
Comunidade Agrícola de Higienópolis	23.79	0.99	0.32
Alto Bela Vista	29.18	1.22	0.35
Parque São Sebastião	14.75	0.61	0.18
Fazenda Botafogo/Margem da Linha	31.56	1.32	0.32
Pantanal	25.16	1.05	0.25
Catumbi	79.17	1.10	0.26
Morro União	27.05	1.13	0.28
Chico Mendes (Morro do Chapadão)	29.46	1.23	0.35
Morro do Cariri	23.32	0.97	0.33
Parque Alvorada	11.34	0.47	0.14
Santa Terezinha	10.89	0.45	0.14
Borel	25.29	1.05	0.29
Parque João Paulo II	20.36	0.85	0.26
Vidigal	17.54	0.73	0.21
Parque Vila Isabel	25.04	1.04	0.29
Pavão-Pavãozinho	26.99	1.12	0.25
Muzema	10.40	0.43	0.13
Parque Tiradentes	23.26	0.97	0.31
Total	933.23	0.85	0.25

Source: Own elaboration from CENTROCLIMA/COPPE/UFF/SMAC. Relatório Final do Projeto Diagnóstico Preliminar de Resíduos Sólidos da Cidade do Rio de Janeiro; 2015 [6].

6.3 Final considerations

In the 2010 census [5], subnormal agglomerates corresponded to 6% (11.4 million) of the Brazilian population, spread over only 323 municipalities (6% of the total number).

Table 6.10 Volume of total waste generated per household by material type—Catumbi.

Material type	Weight (kg)	
	Total	**Per household**
Plastics	227.95	3.17
Tetra Pak	95.13	1.32
Paper	91.49	1.27
Glass	131.77	1.83
Organics	2865.06	39.79
Metal	79.17	1.10
Total	3490.56	48.48

Source: Own elaboration from CENTROCLIMA/COPPE/UFF/SMAC. Relatório Final do Projeto Diagnóstico Preliminar de Resíduos Sólidos da Cidade do Rio de Janeiro; 2015 [6].

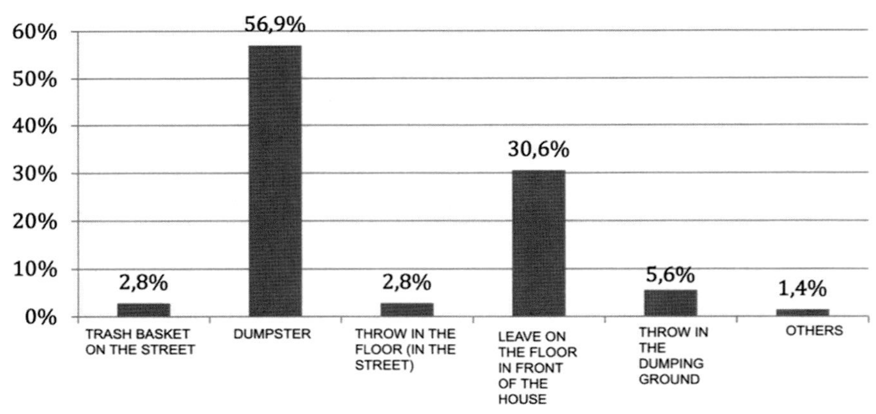

Figure 6.8 Distribution of households by waste disposal's type. *Own elaboration from CENTROCLIMA/COPPE/UFF/SMAC. Relatório Final do Projeto Diagnóstico Preliminar de Resíduos Sólidos da Cidade do Rio de Janeiro; 2015 [6].*

Almost half of this contingent was in the Southeast, with emphasis on the states of São Paulo (2.7 million) and Rio de Janeiro (2.0 million).

In a nutshell, these are the main results of the study carried on by [6]:

1. It has been found that much of what will end up in dumps or landfills could be recycled. In 2012 Brazil produced almost 63 million tonnes of household solid waste, which is just below the world per capita average, that is, 1 kg per inhabitant per day. This average, however, hides immense inequality, with which data are precarious: for example, in the recent study conducted by Comlurb, at Morro Dona Marta, in Rio de Janeiro, reveals that in 2012 its residents generated 0.53 kg, individually, almost half therefore of the national average.

2. In the 2010 census, subnormal agglomerations accounted for 6% (11.4 million) of the Brazilian population and distributed in only 323 municipalities (6% of the total number). Almost half of this contingent was in the Southeast, especially the states of São Paulo (2.7 million) and Rio de Janeiro (2.0 million).
3. The total waste generated in the Catumbi sample was approximately 3491 kg/month, with a value per household of about 48.5 kg/month and a daily value per household of 1.6 kg. It was concluded that considering the average of 4 residents per household, per capita generation obtained corresponds to 0.4 kg/day. This number is not significant for the Catumbi favela in relation to per capita waste generation, according to [13]—reaching 1.2 kg per person per day. It should be noted that the consumption profile and income pattern considerably reduce the amount of packaging increasing the organic generation.
4. Catumbi presented itself through field research as the third largest waste generator in the total monthly, with 3490.56 kg; second, Rio das Pedras (4034.60 kg); and Rocinha leading with 5568.42 kg of discarded materials per month.

From all that has been exposed, we may say that the fundamental issue of solid waste calls for a management model that prioritizes the review of generating processes, starting with consumption habits, and that likewise, it has a state-of-the-art policy in recycling. Since excessive use of new materials will result in a market failure, the garbage problem is an example of negative externality, in the face of environmental pollution, impacting its assimilative capacity above its capacity for regeneration, and causing social costs not compensated [14].

If, on the one hand, the course of human activity is considered impossible without the generation of waste, on the other hand, such waste should be disposed of correctly. Furthermore, the amount of waste produced is directly associated with the degree of economic development in a country. The richer a nation is, the more garbage it produces. The persistent dynamic relationship of the problem of managing solid waste between individuals and the State, registered the perception of the need for rules that would guarantee the fundamental rights of the human being, as an attempt to stop their generation. Thus in 2010 Federal Law 12,305 [15] was published, instituting the National Solid Waste Policy, which establishes guidelines for integrated management and solid waste management and defines the responsibilities of generators and public authorities.

There is a great difficulty in characterizing solid waste management in Brazilian cities, since most of them do not have weighing scales for waste, open pit dumps—landfills—and end up making the evolution of this issue unfeasible in Brazil. The sustainable management of solid waste presupposes an approach based on the principle of the 3Rs: (amount of waste generated) reduction; (direct) reuse of products before disposal; and materials recycling. These are important concepts that must be assimilated, practiced, and propagated.

Moreover, much of what ends up in landfills or landfills could be recycled. In 2012, Brazil produced almost 63 million tonnes of solid household waste, standing just below the world average per capita, that is, 1 kg per inhabitant per day. This average, however, hides immense inequality, for which the data is precarious: for example, a study conducted by Comlurb [16], in Morro Dona Marta, in Rio de Janeiro, reveals that in 2012 its residents generated 0, 53 kg, individually, almost half therefore of the national average.

We may conclude that in fact the promotion of environmental education at all levels of education and mainly public awareness for the preservation of the environment is a fundamental policy. The management of urban solid waste must also be present in organs, institutions, associations, and the favelas as a whole. It is necessary to be fully involved in environmental education and learning programs, with an emphasis on raising awareness to reduce the generation of waste—since nongeneration is impossible; and mainly for the correct separation of recyclable waste, which will contribute to the correct final destination, aiming to minimize environmental impacts in favor of a better life quality. Historically, this situation points out that the mistakes made in dealing with the issue of the generation and destination of waste persists, contributing to greater difficulties.

References

[1] Alvito M. Dos Parques-proletários ao Favela-Bairro: as políticas públicas nas favelas do Rio de Janeiro. In: ALVITO, ZALUAR (org.), editors. Um século de favela. Rio de Janeiro: FGV; 1998.

[2] Valladares LdP. Passa-se uma casa: análise do programa de remoção de favelas do Rio de Janeiro. Rio de Janeiro, Zahar; 1978.

[3] Birman P. Favela é comunidade? Em: SILVA, Luiz Antônio Machado da (org). Vida sob cerco: violência e rotina nas favelas do Rio de Janeiro. Rio de Janeiro, Nova Fronteira; 2008.

[4] Prefeitura da Cidade do Rio de Janeiro. Armazém de dados. http://www.armazemdedados.rio.rj.gov.br/ [accessed 12.12.13].

[5] IBGE Censo Demográfico 2010—Aglomerados Subnormais—primeiros resultados, Rio de Janeiro, RJ: IBGE; 2011.

[6] CENTROCLIMA/COPPE/UFF/SMAC. Relatório Final do Projeto Diagnóstico Preliminar de Resíduos Sólidos da Cidade do Rio de Janeiro; 2015.

[7] SEBRAE. Serviço Brasileiro de Apoio às Micro e Pequenas Empresas. http://www.sebrae.com.br [accessed 14.01.14].

[8] FEIRA FEITA. http://www.feirafeita.com.br [accessed 14.01.14].

[9] Hortifruti. http://www.hortifruti.com.br [accessed 14.01.14].

[10] Harada TM, Philippi ST. Padronização de medidas usuais de consumo de frutas das regiões Centro-Oeste e Nordeste Brasileiras. FSP/USP; 2006.

[11] Moura P, Honaiser A, Bolognini M. Avaliação do Índice de resto ingestão e sobras em unidade de alimentação e nutrição do colégio agrícola de Guarapuava (PR). Universidade Estadual do Centro-Oeste/Unicentro; 2010.

[12] Plastikero. http://www.plastikero.com.br/ [accessed 13.01.13].

[13] World Bank. What a waste. A global review of solid waste management. http://web.worldbank.org/wbsite/external/topics/exturbandevelopment/0,contentmdk:23172887~pagepk:210058~pipk:210062~thesitepk:337178,00.html; 2012 [accessed 19.11.13].

[14] Valerio D, Carestiato T, Cohen C. Redução da Geração de Resíduo Sólidos: Uma Abordagem Econômica. http://www.anpec.org.br/encontro2008/atrigos/200807211417570pdf [accessed 12.10.13].

[15] BRASIL. Federal Law nº 12.305, of August 2, 2010. Establishes the National Solid Waste Policy. http://www.planalto.gov.br/ccivil_03/_ato2007-2010/2010/lei/l12305.htm [accessed 14.01.14].
[16] Comlurb. Comunicação pessoal a partir de Análise Gravimétrica dos resíduos sólidos urbanos recolhidos pela COMLURB para o Município do Rio de Janeiro—*2012*, Paulo Jardim, COMLURB, sobre Coleta Domiciliar, Lixo Público e Entulho; 2014.

CHAPTER 7

Transesterification process of biodiesel production from nonedible vegetable oil sources using catalysts from waste sources

7.1 Introduction

Highly wavering crude oil prices with depleting oil sources affect many countries' economies, especially oil importers like India, placing an intense encumbrance on their foreign exchange [1]. One way of solving this problem is to identify alternative energy sources, which preferred to be renewable. Biomass sources, mainly vegetable oils, have attracted much attention toward alternative fuels. Several researchers have reported that using pure vegetable oil in existing diesel engines can cause various engine-related issues, such as extreme deposits, injector coking, and piston ring sticking, due to its low volatility and high viscosity as a result of the need for oils with lower emissions, higher volatility, and lower viscosity. Biodiesel is one of the most efficient alternative fuels for diesel engines since it is made from renewable sources and is biodegradable, nontoxic, and environmental friendly [2].

Generally, farmers' world burns their agricultural residues as they do not have a way to recycle them efficiently. This incinerating of the wastes and residues also contributes to global warming as they release a massive amount of carbon dioxide into the atmosphere. Greenhouse gas emissions are a significant concern for ozone layer deficiency to cause severe effects, including glaciers' melting and temperature rise. The global energy demand is also higher than ever, sustainably requiring an adaptive way with environmental concern. A suitable alternative will be the collection of these agricultural residues and using them for efficient biodiesel production.

Apart from being renewable, biomass-based fuels are highly economical in maintenance and production. Moreover, a significant percentage of industrial wastes and by-products can also be employed for biodiesel production. Currently, incineration, landfills, and waste treatment are focused on by-product management in industries.

This chapter deals with biodiesel production using synthetically derived heterogeneous catalysts, transesterification mechanism, impacts of catalysts on kinetics, variable methodologies of production, and techno-economic analysis (TEA).

A Thermo-Economic Approach to Energy From Waste
DOI: https://doi.org/10.1016/B978-0-12-824357-2.00002-4

7.2 Biodiesel production as an alternative source of energy

Biodiesel production requires a feedstock in the form of biological material, usually vegetable oilseeds or other biological residues. Due to the less sulfur content and renewable nature of vegetable oil is preferred as an alternative fuel. During combustion, sulfur dioxide emissions from vegetable oil are negligible, and toxic emissions are minimal. In addition, vegetable oil is carbon neutral in general; the carbon dioxide released during combustion is absorbed and used for photosynthesis. Thus vegetable oil is biodegradable, safe to store, and transport and has no adverse effects on the environment or human health. However, due to the lower volatility and higher viscosity, vegetable oils in their raw form cause operational and reliability issues in direct injection diesel engines. Injector clogging, piston ring sticking, gum deposits, and lubricating oil thickening were symptoms of long-term operations. As a result, oils with lower viscosity, higher volatility, and lower emissions, such as biodiesel, are in demand.

Biodiesel is a fuel made up of a mixture of mono-alkyl esters of long-chain fatty acids that meets the ASTM D6751 specifications and is extracted from animal fats or vegetable oils. Biodiesel exhibits similar properties and characteristics to that of conventional diesel, as given in Table 7.1 as mentioned by Lotero et al., [3]. Therefore it can be used directly or in combination with traditional diesel engines without requiring extensive engine modifications [4]. Besides, biodiesel offers benefits over petroleum-based diesel, such as (1) a high flash point and a high cetane number, indicating safer and efficient performance,

Table 7.1 Maximum allowable levels in diesel and biodiesel as regulated by the American Society for Testing and Materials (ASTM).

Property	Unit	Biodiesel	Conventional diesel
Standard		ASTM D6751	ASTM D975
Composition		FAME (C12–C22)	HC (C10–C21)
Kinetic viscosity at 40°C	mm^2/s	1.9–6.0	1.9–4.1
Boiling point	°C	182–338	188–343
Cloud point	°C	−3 to 12	−15 to 5
Flash point	°C	100–170	60–80
pour point	°C	−15 to 16	−35 to −15
Hydrogen	wt.%	12	13
Carbon	wt.%	77	87
Sulfur	wt.%	0.05	0.05
Oxygen	wt.%	11	0
Water	vol.%	0.05	0.05
AFR		13.8	15
Centane number (ignition quality)		48–60	40–55
BOCLE scuff	g	>7000	3600
HFRR	μm	314	65
Life-cycle energy balance (energy units produced per unit energy consumed)		3.2/1	0.83/1

FAMEs, Fatty acid methyl esters; *HC* – HydroCarbon; *AFR*, Air/fuel ratio; *HFRR*, High frequency reciprocating rig; *BOCLE* - Ball on cylinder lubricity evaluator.

(2) higher lubricity decreases the need for engine part replacements frequently and prolongs the life of the engine, and (3) the presence of oxygen (~10%) reduces CO and hydrocarbon emissions in biodiesel and improves combustion.

Hence, to lower operational issues with vegetable oils and use renewable energy sources, biodiesel can be an excellent alternative fuel.

7.3 Transesterification: reaction and mechanism

There are several methods of utilizing biomass for the production of energy [5]. Among them, blending, pyrolysis, emulsification, and transesterification were the most commonly used methods for reducing oil viscosity. The simplest of them is the production of biodiesel using pyrolysis. Pyrolysis is a process where the chemical compound is decomposed at a higher temperature to yield different compounds. Pyrolysis has a low processing cost, but it is energy-intensive as it requires high temperature for the production process. Microemulsification resulted in reduced fuel viscosity, but the calorific value of fuel is lower, and hence, this method is not widely used. The transesterification method yields a product with high energy content and higher cetane energy. Transesterification is one of the most commonly adaptive methods with certain modifications to make it a promising technology, though this method requires multiple steps in downstream processing for purification. Another method is the direct blending of vegetable oils with diesel for obtaining biodiesel. The portability is high for this method, but the produced fuel is less volatile and less reactive. So, this method is also not widely used for large-scale applications.

A transesterification reaction is extensively used to produce biodiesel. Oil in the presence of a catalyst reacts with alcohol to produce glycerol and mono-alkyl esters, isolated and purified. Similar to hydrolysis, it is the method of removing alcohol from an ester by replacing it with another alcohol. The reaction mechanism is represented as follows:

$$\begin{array}{lccll} CH_2-O-C(=O)-R_1 & & & CH_3-O-C(=O)-R_1 & \\ | & & & & CH_2-OH \\ CH-O-C(=O)-R_2 & +\ 3\,CH_3OH & \xrightarrow[\text{(Catalyst)}]{} & CH_3-O-C(=O)-R_2\ + & CH-OH \\ | & & & & CH_2-OH \\ CH_2-O-C(=O)-R_3 & & & CH_3-O-C(=O)-R_3 & \\ \text{triglyceride} & \text{methanol} & & \text{mixture of fatty esters} & \text{glycerol} \end{array}$$

The overall process typically consists of three reversible reactions that occur consecutively as given.

$$\text{triglyceride} + R'OH \Leftrightarrow \text{diglyceride} + RCOOR'$$

$$\text{diglyceride} + R'OH \Leftrightarrow \text{monoglyceride} + RCOOR'$$

$$\text{monoglyceride} + R'OH \Leftrightarrow \text{glycerol} + RCOOR'$$

Alcohol and oil have a stoichiometric relation of 3:1. Among the alcohols, such as methanol, ethanol, propanol, and butanol, methanol possesses a proper viscosity, boiling point, and a high cetane number. However, an enormous amount of alcohol is often used to enhance the reaction to obtain the desired product.

7.4 Catalysts

Most biodiesel production from vegetable oil sources or animal fats uses homogeneous catalysts, such as sulfuric acid, hydrochloric acid, sodium hydroxide, sodium methoxide, and potassium hydroxide. Homogeneous acid catalysts afford higher yield but needs longer reaction time and higher reaction temperature for complete conversion; as well, these catalysts are corrosive to the reactors. Although a homogeneous basic catalyst requires less reaction time and mild temperature, the separation of catalyst from the product is a technical glitch. Furthermore, a tremendous amount of wastewater is generated to meet biodiesel's stipulated quality [6]. Besides, the catalyst is consumed to form soap due to the feedstocks' high free fatty acid (FFA) content. Hence, to overcome these issues, heterogeneous catalysts have been developed. Heterogeneous catalysts replace the conventional homogeneous catalysts mainly due to environmental constraints and simplifications in the existing processes [7].

7.4.1 Chemical catalysts

Chemical catalysts for biodiesel production can be classified as homogeneous catalysts, heterogeneous catalysts, and nanocatalysts.

7.4.1.1 Homogeneous catalysts

Homogenous catalysts are advantageous as they offer a high reaction rate despite being highly selective [8]. Generally, alkalis, such as oxides of sodium and potassium, have been applied as chemical catalysts. These catalysts are selected for the process based on their solubility with the solvent used in the production. However, the recovery of products becomes complicated in homogenous alkaline catalysts as they tend to form soap by the saponification of esters. Several reports suggest the usage of both alkalis and acids for biodiesel production [9–11]. In such processes, the acid is used initially for esterification, followed by transesterification with alkaline catalysts. Acids alone can also be used as a catalyst, such as sulfuric acid, phosphoric acid, and hydrochloric acids, to yield a good product. Generally, methanol is preferred as alcohol because it provides better recovery of ester [12]. The catalyst effects can be improved by introducing an external stirrer for adequate dispersion or other energy transfer [13].

7.4.1.2 Heterogeneous catalysts

Some vegetation sources contain high FFA. In such cases, the homogenous catalysts are not able to render their part effectively. Also, homogenous catalysts generate more by-products than heterogeneous catalysts [14]. Heterogeneous catalysts have been adopted increasingly to make the process eco-friendly, economical, and faster by easier recovery of the products [15]. Heterogeneous catalysts include the oxides of alkaline earth metals [16], transition metals, zeolites, other mixed metals [17], etc. The loss of catalysts in the alcohols is negligible during the transesterification reaction because of their insolubility. Besides, heterogeneous catalysts from the reaction mixture could be easily separated by filtration and then reused. The use of a heterogeneous catalyst in the development of biodiesel could decrease the separation and purification steps, promoting commercialization economically. The classification of solid base catalysts is listed in Table 7.2.

Table 7.2 Classification of different heterogeneous basic catalysts.

	Types of the heterogeneous basic catalyst
Zeolites	Alkali ion-supported/ion-exchanged zeolite, TiO_2, ZrO_2, ZnO, ThO_2
Single component metal oxides	Alkaline earth oxide
	Alkali metal oxide
	Rare earth oxides
Clay minerals	Sepiolite
	Hydrotalcite
	Crysotile
Nonoxides	Alkaline alkoxide
	Guanidine containing catalyst
	Alkaline carbonate
Supported alkali metal (or alkaline earth material)	Alkali metal ions on alumina
	Alkali metal on alkaline earth oxides
	Alkali metal ions on silica
	Alkali metals and alkali metal hydroxides on alumina

Catalyst preparation methods are highly complex, and each catalyst can be synthesized in various ways. In most cases, preparation entails a series of steps. Coprecipitation, impregnation, drying, calcination, and activation are all steps to make supported metal oxide catalysts. Similarly, gel precipitation, crystallization, washing, ion exchange, and drying were used to prepare zeolite catalysts. Heterogeneous catalysts' previous history determines their properties. Catalyst preparation's most crucial phases may be:

1. First precursory solid or primary solid preparation that would be associated with all of the useful components. For instance, impregnation or coprecipitation.
2. Preparation of the catalyst precursor from the obtained primary solid (heat treatment).
3. Activation of the precursor to produce the active catalyst (i.e., calcination).
4. Forming operations by drying.

7.4.1.3 Preparation of natural derived heterogeneous catalyst

The eggshells and white bivalve clamshells (WBCS) are thoroughly washed in water and rinsed twice with purified water to remove an unnecessary material adhered to their surface. Fig. 7.1 depicts a white bivalve clam shell after being washed with distilled water. The washed shells are dried for 24 hours in a hot air oven at 105°C. In eggshells, the dried shells are crushed into tiny fragments and then calcined at 900°C for 2.5 hours under static air conditions in a muffle furnace. In WBCS, calcination is done at 900°C for 4 hours to make active CaO particles from calcium species in the shell. With CO_2 and water, alkali earth metal oxides have a high tendency to react actively. To prevent the CO_2 and moisture in the air from reacting, the calcined shells (eggshells CaO and WBCS CaO) will be ground to a fine powder and stored in an airtight closed vessel.

Figure 7.1 Image of the washed white bivalve clam shells.

To improve the catalytic activity of the catalysts, eggshell CaO and WBCS CaO are subjected to various treatments. The solid particles from eggshell CaO and WBCS CaO are filtered and dried in a hot air oven at 120°C overnight after being refluxed in water at 60°C for 6 hours. The solid product is then dehydrated by re-calcining it at 600°C for 3 hours to convert it from hydroxide to oxide. As a result of the calcination, hydration, and dehydration procedures, the shells produce a highly active CaO surface [18–22].

Functionalization of catalysts is performed to improve the basic catalytic activity with different mass ratios of potassium fluoride (KF) at 0.15, 0.25, and 0.35 by impregnating CaO with an aqueous KF solution. The prepared catalyst was calcined for 4 hours under static air conditions in a muffle furnace at 600°C. Impregnation CaO can also be executed with aqueous potassium hydroxide (KOH) solution, with different mass ratios of 0.25, 0.35, and 0.45, then calcined for 4 hours under static air conditions in a muffle furnace at 600°C.

7.4.1.4 Nanocatalysts

Nanocatalysts are those that combine the benefits of both the homogenous and heterogeneous catalysts. They provide a very high yield with high reaction rates, hence solving the problem of catalysis. Research is being done extensively on these materials, which acquire an increased number of nanoreaction sites and offer a low alcohol/oil separation ratio. Nanocatalysts of CaO yielded a higher recovery of biodiesel, which is commendable [23,24].

7.4.2 Biochemical catalysts

Chemical catalysts usually have some demerits, such as less biodiesel recovery, severe conditions requirement, and high cost of the catalysts. Usually, these biochemical catalysts are enzymes that are used due to their compatibility and biodegradability. These are greener catalysts for contributing to a sustainable environment and have the advantages of high-end properties, such as regioselectivity and specificity, leading to efficient biodiesel production and higher reaction rates in the transesterification reaction. Generally, these enzymes are immobilized by adsorption, encapsulation, or other techniques to improve all their properties [25]. Enzyme catalysts lead to lighter operating conditions, which reduces the operating cost with higher production [26]. The usage of biochemical catalysts and enzymatic transesterification has a promising future, indeed looking at the world's current trends focusing on sustainability and greener materials.

7.4.3 Impact on kinetics of transesterification and modeling

The change in moles of a component per unit volume of the reaction mixture is known as the reaction rate. The reaction rate is negative for the reactant component, while for the product component, it is positive. The rate refers to a differential volume of the reaction mixture and must be a local, or point, value. If the rate is constant throughout the reactor's volume, the concentrations and temperature must be uniform. Otherwise, the rate will vary at each point of the reaction volume.

In general, the reaction rate is given in Eq. (7.1)

$$aA + bB \rightarrow cC + dD \tag{7.1}$$

It may be written as,

$$r = k[A]^a[B]^b \tag{7.2}$$

The order of the reaction with respect to A and B is a and b, and the proportionality constant k is known as the specific rate constant. The specific rate constant k in Eq. (7.2) generally depends on temperature. The rate constant also varies with other variables but is significant, such as the nature and extent of the catalytic substance's surface, concentration, and nature of the components.

7.4.3.1 Determination of kinetic parameters in a batch process

Transesterification is a catalyst-assisted reaction in which fat or oil reacts with alcohol to produce esters and glycerol. The catalyst is assumed to be used in sufficient oil to attain the reaction equilibrium shift to form fatty acid methyl esters (FAMEs). Thus the catalyst's change in concentration during the reaction becomes negligible and not considered the reverse reaction.

The transesterification is assumed as a single-step reaction, and the rate law for forward reaction can be expressed by Eq. (7.3)

$$-r_A = \frac{-d[\mathrm{TG}]}{dt} = k^1[\mathrm{TG}][\mathrm{ROH}]^3 \tag{7.3}$$

where [TG] is the concentration of triglycerides, [ROH] is the concentration of methanol, t is the time, r_A is the reaction rate, and k^1 is the equilibrium rate constant. A second-order reaction rate law regulated the above overall reaction. The shift in methanol concentration can be considered constant during the reaction because of the high molar ratio of methanol to oil. This means that the excess methanol does not alter the reaction's order, and it behaves like a first-order chemical reaction. Finally, the rate expression can be written as given in Eq. (7.4):

$$-r_A = \frac{-d[\mathrm{TG}]}{dt} = k[\mathrm{TG}] \tag{7.4}$$

where k is known as the modified rate constant and $k = k^1\ [\mathrm{ROH}]^3$

For solving, it is assumed that at time $t=0$, the initial triglyceride concentration to be $[TG_0]$, and at the time t, the triglyceride concentration to be $[TG_t]$.

The integration of Eq. (7.5) yielded Eq. (7.6)

$$\int_{[TG_0]}^{[TG_t]} \frac{-d[TG]}{dt} = \int_0^t k dt \tag{7.5}$$

$$\ln[TG_0] - \ln[TG_t] = kt \tag{7.6}$$

$$\ln \frac{[TG_0]}{[TG_t]} = kt$$

From mass balance, Eq. (7.7) can be evaluated as

$$[TG_t] = [TG_0] - [TG_0][X_A]$$

$$[TG_t] = [TG_0](1 - X_A)$$

$$\frac{[TG_t]}{[TG_0]} = (1 - X_A) \tag{7.7}$$

Taking natural log on both sides, Eq. (7.7) can be rewritten as Eq. (7.8)

$$\ln \frac{[TG_t]}{[TG_0]} = \ln(1 - X_A) \tag{7.8}$$

On solving Eqs. (7.6) and (7.8), to obtain Eq. (7.9), which can be applied to determine the rate constant.

$$-\ln(1 - X_A) = kt \tag{7.9}$$

7.4.3.2 Modeling of batch reactor design

The rate equation can be expressed in Eq. (7.4) by assuming the reaction kinetics to be pseudo-first-order. In a batch reactor, the composition is uniform throughout at any instant of time; the material balance accounting may be made about the whole reactor. Taking material balance for the whole reactor as follows in Eq. (7.10)

$$\text{Rate of input} = \text{rate of output} + \text{rate of disappearance} + \text{rate of accumulation} \tag{7.10}$$

Assuming there is no fluid entered or left the reaction mixture during the reaction, the above Eq. (7.10) can be modified as Eq. (7.11)

$$\text{Rate of disappearance of reactant} = -(\text{rate of accumulation of reactant in the reactor}) \quad (7.11)$$

$$\text{Rate of disappearance of reactant} = -r_A V = kC_{TG} V \quad (7.12)$$

Rate of accumulation of reactant within the reactor:

$$\frac{dN_{TG}}{dt} = \frac{d[N_{TG}(1 - X_{TG})]}{dt} \quad (7.13)$$

On simplification of Eq. (7.13) yielded Eq. (7.14)

$$\frac{dN_{TG}}{dt} = \frac{N_{TG_0}}{C_{TG_0}} \frac{dC_{TG}}{dt} \quad (7.14)$$

Substituting Eqs. (7.12) and (7.14) in Eq. (7.11) and rearranged to obtain Eq. (7.15)

$$kC_{TG} V = -\frac{N_{TG_0}}{C_{TG_0}} \frac{dC_{TG}}{dt} \quad (7.15)$$

$$\frac{dC_{TG}}{dt} = -\frac{KC_{TG_0} V}{N_{TG_0}} dt$$

The integration of Eq. (7.15) at $t = 0$, $[TG] = [TG_0]$ and at $t = t$, $[TG] = [TG_t]$ and on further solving resulted the following Eq. (7.16) to evaluate the final concentration of triglycerides for calculating the yield of biodiesel.

$$\frac{C_{TG}}{C_{TG_0}} = e^{-\frac{KC_{TG_0} V}{N_{TG_0}} t}$$

$$C_{TG} = C_{TG_0} e^{-\frac{KC_{TG_0} \quad V}{N_{TG_0}} t} \quad (7.16)$$

where t is the time (minutes), V is the volume of reaction mixture (L), k is the rate constant (min^{-1}), N_{TG_0} is the initial number of moles of triglycerides (moles), C_{TG_0} is the initial concentration of triglycerides (moles/L), and C_{TG} is the final concentration of triglycerides (moles/L).

7.4.3.3 Modeling for continuous reactor design

The main difference between homogeneous reactions and heterogeneous reactions (fluid-solid) for reactor design calculations is that the solid catalyst mass and W determines the reaction rate for the latter, rather than the reactor volume, V. The reaction rate of substance A in a fluid-solid heterogeneous system is represented as Eq. (7.17)

$$-r_A = \text{mole of A reacted}/(\text{time} * \text{mass of catalyst}) \tag{7.17}$$

In a packed bed reactor, the mass of the solid is often used since the catalyst amount is a critical factor for a reaction rate. Also, there will not be any accumulation in transesterification reaction for the production of biodiesel. It is assumed that there is no radial gradient in temperature, concentration, or reaction rate.

The generalized material balance equation on the reactants over catalyst weight results in Eq. (7.18)

$$\text{In} - \text{out} + \text{generation} = \text{accumulation}$$

$$F_A(W) - F_A(W + \Delta W) + -r_A \Delta W = 0 \tag{7.18}$$

The dimensions of the generic term in Eq. (7.18) can be shown in Eq. (7.19) as similar as the same dimension of the molar flow rate F_A

$$-r_A \Delta W = \frac{\text{moles}A}{(\text{time})(\text{mass of catalyst})} \times (\text{mass of catalyst}) = \frac{\text{moles}A}{\text{time}} \tag{7.19}$$

After dividing by ΔW and taking the limit as $\Delta W \rightarrow 0$, the differential form of the equation can arrive for a packed bed reactor as Eq. (7.20)

$$\frac{dF_A}{dW} = r_A \tag{7.20}$$

In which, F_A can be substituted as $C_A m^1$ and r_A as kC_A, then Eq. (7.20) can be modified as Eq. (7.21),

$$dW = \frac{-dC_A m^1}{kC_A} \tag{7.21}$$

The integration of Eq. (7.21) at $t = 0$, $[TG] = [TG_0]$ and at $t = t$, $[TG] = [TG_t]$ and on further solving resulted the following Eq. (7.22) to evaluate the final concentration of triglycerides for calculating the conversion of biodiesel.

$$C_A = C_{A_0} e^{\frac{-Wk}{m^1}} \quad \text{or} \quad (1 - X) = e^{\frac{-Wk}{m^1}} \tag{7.22}$$

where m^1 is the mass flow rate (g/min), W is the weight of the reactor contents (g), X is the conversion of triglycerides to biodiesel, C_{A_0} is the initial concentration of triglycerides (mol/g of catalyst), C_A is the final concentration of triglycerides (mol/g of catalyst), and k is the rate constant of the reaction (min^{-1}).

7.5 Hydrocarbon feed stocks for biodiesel

7.5.1 Edible oils

Vegetable oil is biodegradable, easy for storage, and transport and has no adverse effects on the environment or human health. For biodiesel processing, edible vegetable oils from various sources, such as rapeseed, sunflower, soybean, cottonseed, palm, olive, and others, have been used. The feedstock, which accounts for 70%–95% of the total production cost of biodiesel, substantially affects production costs [27]. The critical issue for a large-scale biodiesel application is costlier than petroleum-based conventional diesel. Biodiesel's high cost is due to a shortage of feedstocks and a lack of edible oils. Therefore exploring ways to reduce the cost of feedstock plays a vital role in recent biodiesel research.

7.5.2 Nonedible oils

Using feedstocks, including animal fats, nonedible oils, and waste frying oils as feedstocks is an efficient way to minimize biodiesel output. However, since these low-cost products usually contain many FFA, the alkali-catalyzed transesterification process produces lower yields. A base catalyst cannot perform transesterification reactions if the oil's FFA content exceeds 3 wt.% [28]. Nonedible oils from the sources, such as jatropha, mahua, Karanja, neem, and cottonseed, do not compete with food industries. These oils could be considered for biodiesel production as they are relatively cheaper than edible oils and their availability is throughout the year in our country. Most of the meat processing industries release many animal fat and lard as wastes that could be applied to biodiesel production [29]. Other feedstocks include agricultural wastes, bio residues, microalgae, and pretreated waste frying oil, are also employed to reduce the production cost of biodiesel [30,31].

7.6 Various novel technologies for biodiesel production

In recent days several technologies are being increasingly used to produce biodiesel to improve the performance for making it more efficient. The traditional method involves heating and stirring, which are energy-intensive processes. One of the disadvantages of the conventional method is that it requires oil extraction from the biomass before transesterification to form biodiesel. The process of solvent extraction is expensive and has several safety issues. Also, there is a significant wastage of biomass in the conventional process as mechanical methods yield less oil. These subtle challenges lead to severely inefficient processes if not governed properly. Hence, a requirement of novel technologies arises that could solve traditional processes, such as solvent extraction and mechanical operations. The most promising techniques, such as ultrasonic-assisted biodiesel production, microwave-assisted transesterification, micro reactive transesterification, and supercritical technology of biodiesel production. These techniques are analyzed in detail in this section.

7.6.1 Ultrasonic-assisted biodiesel production

The conventional method of transesterification reaction is very slow, and it needs catalysts to enhance the reaction rate. In recent years ultrasound has come up as a sustainable solution for catalysis of this transesterification reaction to make the process cost-effective and result in a high yield [32]. Ultrasound is a form of electromagnetic wave with higher frequencies that provides the energy required for mixing the liquids in the biodiesel reactors. Hence, the required activation energy is reached easily [33]. This eliminates the slow mixing of the liquids, which is an influencing factor for contributing to higher biodiesel yield.

The mechanism of ultrasound aids the formation of cavities due to pressure differences by a process called cavitation [34]. This leads to an enhanced reaction rate to yield a higher product with lesser reaction time [35–37] using enzyme catalyst. Furthermore, under the same reaction conditions. In the presence of ultrasound, heterogeneous base-catalyzed reactions for methanolysis are more rapid rather than the conventional mechanical stirring. This may be because ultrasound causes changes in the surface pattern of solid catalysts, resulting in new catalytically active surface area and long-term catalytic activity [38].

Adewale et al. have found a 96.8% yield in their experiment with lard with an ultrasonic-assisted mechanism which was enzyme-catalyzed. They used *Candida antarctica* lipase B to give this high yield within 20 minutes of reaction. The conventional nonultrasound-assisted method takes nearly 22 hours to produce the same yield. Similarly, with coconut oil as raw material, Tupufia et al. have achieved a yield of 92% for ultrasound-assisted conversion. They used Novozyme-435 as a catalyst to curb the side reactions in this process. This yield conventionally takes almost 50 hours to be achieved. Hence it can be inferred that ultrasonic waves can catalyze this esterification reaction effectively for high yields with lesser reaction time. Suresh et al. [39] used ultrasonic treatment to produce biodiesel from residues of the meat industry. The optimal parameters found using RSM were 35.36% of United States time, 2.07% of CuO catalyst with methanol to PTO ratio of 2.07 wt.%. Using the determined parameters, a yield of 97.82% was obtained.

The use of phase transfer catalysts, process variable optimization, process kinetic model development, continuous process implementation, the design of novel ultrasonic reactors, and in situ ultrasound in transesterification is the emerging methods for lowering biodiesel production costs. To achieve a higher degree of TGA conversion with uniform product quality and the overall economic effect, low retention time, lower reactor volume, and lower production cost were considered to be achieved efficiently by the continuous process of ultrasonic transesterification, such as ultrasonic flow reactors. Future research should concentrate on creating new ultrasonic reactor models that are more efficient than those currently in use. An excellent way to build novel, more powerful ultrasonic reactors is to combine flow reactors with ultrasound action, such as loop reactors, sequence units of tubular reactors, and pipe reactors [40].

7.6.2 Micro reactive transesterification

Microreactive transesterification is an efficient alternative to the conventional technologies for biodiesel production as it ensures proper mixing of the reactive components with reduced reaction time [41]. Reaction time is a critical constraint in the transesterification process as they are highly energy intensive, and high reaction time requires an enormous amount of power for production. Also, the yield is lesser in these traditional methods as they do not ensure proper mixing of the liquids. Microreactors come in different shapes and sizes and work on varied mechanisms. They provide rapid reaction and enable proper mixing because of their shorter distance of diffusion [42]. Membrane microreactors are used in places where high selectivity is desired. They efficiently control the mixing of liquids and provide a high surface area for the reaction. Based on the feedstock and the solvents, different membranes should be used in the microreactor. Membrane microreactors are based on the rate of mass transfer, which changes the membranes' selectivity to influence the reaction parameters. Different types of membrane reactors have been studied for biodiesel production, such as ceramic carbon membrane reactors [43–45].

Microtube microreactors, a type of tubular reactor, due to the presence of microtubes, provide efficient heat dissipation during the process along with a short distance of diffusion and hence well suited for efficient mixing of several components in the biodiesel reaction mixture by a capillary mechanism. The transesterification reaction takes more time in a microtube reactor than a conventional batch reactor. To further improve the mixing of the alcohol and the oil in the reactor, microstructured [46] reactors can compact solutions for process intensification, providing higher surface area, the higher surface-to-volume ratio to occur the reaction efficiently [47].

Oscillatory flow reactors, the superimposed oscillatory motion creates a flow pattern with the net flow of process fluid, helps efficient mass transfer, and enhances radial mixing with a minimum Reynolds number. Due to the small-scale footprint, lower pumping and capital costs, and easier control helps to obtain high residence times with reduced methanol-to-oil ratio. Further, the oscillatory flow reactors were usually constructed with a lower length to diameter ratio, which inevitably helps get the better economy in biodiesel production [48].

Microfluidic technology, which uses a network of microchannels to drive and store droplets and tiny aliquots of liquids, has revolutionized separation procedures and liquid-based reactions. Since droplets have a high surface-to-volume ratio, they increase the material interface between the reagents and are thus more stable for transesterification [49]. It has been discovered that circulation inside the segmented liquid greatly improves mass transfer between the boundaries of the immiscible liquid within a microchannel. To make the microreactor more commercially accessible, more work needs to be done on the surface design, inlet and outlet methods, channel characteristics, and suitable environment control [50].

7.6.3 Microwave-assisted biodiesel production

Microwaves are a demonstrated method of enhancing chemical reactions as they efficiently increase the energy of the reaction by delivering energy directly to the reactants. Reducing the reaction time can aid the economy of the product by making the heat transfer phenomena more effective, and hence, fewer resources are required [51]. Like ultrasound waves, microwaves are also a type of electromagnetic wave. The basic idea of using microwaves for enhancing biodiesel production is the lower power requirement of the EM waves. They impart their energy to the reaction mix and reduce the reaction time from several hours to several minutes. Microwaves come with the demerit of being costlier [52] than the conventional methods, but an optimal setting of process parameters can reduce costs. Microwaves also work on similar principles, such as the ultrasound waves to create cavitation and impart their energies. However, the increase in microwave strength can cause chemical decomposition of reactants, leading to adverse effects in biodiesel production.

Similarly, the excess temperature may also affect the performance of the process during the reaction. This microwave-assisted biodiesel production can also be coupled with catalysts to improve the reaction rate, particularly found to be efficient in the case of heterogeneous and enzymatic catalyst systems [53]. Also, the improved performance can be claimed using a hybrid version of microwave and ultrasonic-assisted methods for low energy demand with shorter duration [54–56]. Apart from the reaction temperature, reaction time, and the catalyst, the other parameters that affect the conversion yield include FFA of the oil, moisture, and water content of the oil, mixing intensity, the molar ratio of alcohol and oil, and the use of organic cosolvent. The microwave-assisted methods can be improved greatly using nanocatalysts as $MgO/MgFe_2O_4$ by increasing the pore volume and high effective area for the transesterification reaction [57]. Lowering the heating energy requirement reduces fuel cost to a great extent, which could be achieved through a continuous operation rather than a batch process. The continuous operation can achieve higher productivity with lower alcohol requirement and residence time and better safety concerns than batch flow. Hence, it is possible to make the process economically viable through continuous microwave-based transesterification of biodiesel production [58].

7.6.4 Reactive distilled transesterification

The reactive distilled (RD) column is designed to combine the reaction and separation processes in a single column, enhancing mass transfer and simplifying the process flowsheet and operation. High conversion rate, improvement of reaction selectivity, lower energy consumption, solvents' removal in the separation stage, and azeotropes voidance were the few beneficial features achieved through this method [59,60].

In a jacketed RD column, continuous transesterification of feedstock oil with methanol can be possibly developed using heterogeneous catalysts, such as eggshells derived CaO and WBCS CaO. The methanol recirculation in the reactive region and easier product separation are the most significant advantages of the RD method using heterogeneous catalysts. The schematic representation of the model prototype of the RD system is shown in Fig. 7.2. The device consists of a long-packed bed reactor column made up of glass with a jacket provision for hot water circulation to maintain a controlled temperature at 65°C in the column. Up to a certain height of the column, it is filled with glass beads of 1–2 mm diameter, and the remaining portion of the column was filled with catalyst particles, respectively. The bottom end of the column is attached to the middle neck of a three-necked vessel, which could be served as a reboiler. A water-cooled condenser is attached to the top of the column to recover the unreacted reactants and methanol, which are then recirculated back into the reactive zone. The reboiler temperature was kept constant at 65°C during the operation by using a constant temperature water bath for providing heat for the reboiler. The reactive mixture of methanol and oil is thoroughly mixed primarily at a specific ratio

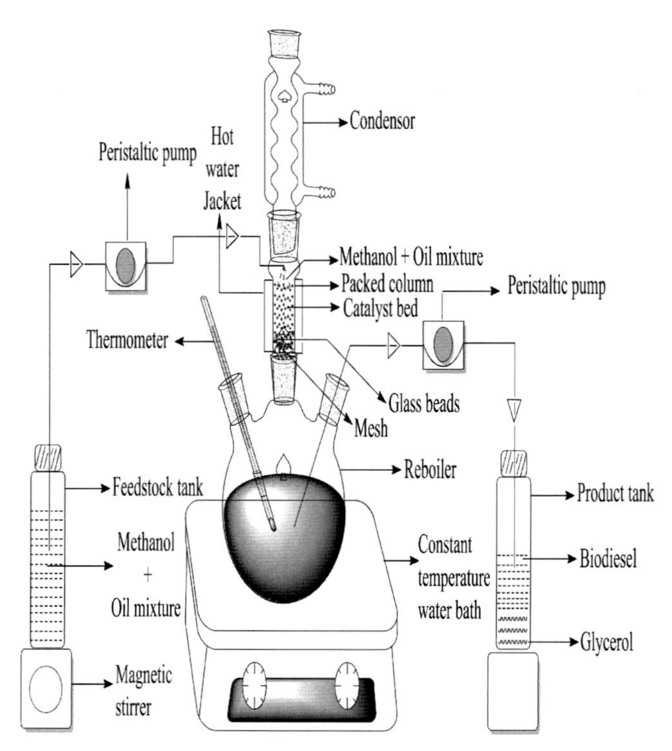

Figure 7.2 Schematic diagram represents the experimental setup of the reactive distillation system.

using a magnetic stirrer, continuously fed into the catalytic section. The majority of the methanol was used in the reaction, and the unreacted methanol, as well as any excess methanol, slowly flowed down to the bottom section, where the reboiler promotes to vaporize the methanol, which is then recirculated in the RD column for effective application in the reactive region. The products of FAMEs and glycerol are collected in the reboiler for further purification, suggesting that FAME and glycerol reverse transesterification are minimal since all catalysts are kept the column [61,62].

7.6.5 Supercritical technology of biodiesel production (noncatalytic)

A supercritical substance is above its critical parameters (critical pressure and critical temperature) but not condensed to a solid. Supercritical fluids, such as carbon dioxide (CO_2), Dimethyl formamide (DMF), and methanol (MeOH), as solvents for the transesterification process without the presence of catalysts. These solvents in their supercritical state provide an efficient medium for reaction as suitable to enhance the mass transfer phenomena during the process and increase the reaction rate, which makes this technology a cleaner approach for biodiesel production. The mechanism of this process is that these solvents, upon pressurization at supercritical conditions, provide homogeneity between reactants, hence allowing efficient mixing of the components in the reaction as similar as the energy imparting by ultrasound or microwave techniques [63,64]. High pressure and temperature are required to attain the homogenous conditions between the reagents and the glycerides where the reaction kinetics are favored. This method can lead to efficient biodiesel production without the presence of catalysts. Another positive impact of the use of supercritical fluid (SCF) technology aids the easier isolation of the produced biodiesel from the biodiesel-glycerol mixture, hence leading to cheaper production of biodiesel [65].

This technology can be employed with dual solvent [66] using supercritical CO_2 with methanol as a cosolvent. Recently, ionic liquids as supercritical fluids could be explored for biodiesel production [67]. This technology is energy-intensive and requires extreme conditions to bring the solvent to a supercritical state. Hence, this phenomenon's heat and mass transfer must be improved for sustainable usage of supercritical fluids [68]. Supercritical technology has a glorious future with some challenges, which need to be tackled for the efficient manufacturing of biodiesel.

7.7 Techno-economic analysis of biodiesel production

TEA is a systematic framework used to analyze an investment's economic and technical performance and determine several economic parameters like the time taken for the return of investment and the different costs involved in the production rate of return, and several other factors. The investment in production depends on several factors, such as raw material cost, operating costs, fixed costs, one-time costs, and other miscellaneous factors on which investment depends on the production. In monetary terms, TEA is used to determine the feasibility to find out the economic benefit of the process to the community.

7.7.1 One-time costs

One-time costs are usually involved at the beginning of the project. It includes initial requirements for an investment to make it operational and to generate revenues. For example, the price of the land where the factory has to be laid is a one-time cost. This directly influences the operational costs, as it also decided the extent of transport required to supply raw materials to the production plant and the transportation of the finished product to the markets. Hence, the site for the project must be decided intelligently to manage the one-time costs and minimize the operational costs. One-time costs also include the cost of setting up the factory and the equipment required for production. These are dependent on the manufacturer, the quality of the equipment, and the production capacity. If a high production amount per day is required, then a continuous production mechanism should be used. These will lead to different production costs when compared to batch reactors. Generally, the materials are decided based on minimizing the total equipment cost and the future maintenance costs. The change in equipment cost is shown in the following equation for different chemical engineering plant cost indices of α for the time of construction and β after a particular amount of time.

$$C_{new} = C_{old}\left(\frac{\beta}{\alpha}\right)$$

For biodiesel production, another factor to be considered is the feedstock used. The feedstock should be locally available, and the yield should be high, leading to higher profits. Different feedstocks may be available locally where the plant is to be set up. Still, the investment on production must be selected based on the raw material for the plant to reduce both the raw material cost and the transportation cost if the material is locally available.

7.7.2 Raw material and operating cost

Raw material cost includes the reactants, catalysts, reagents, etc. Suppose the treatment of the biomass is done by chemical reagents, such as acids, alkalis, or Ionic liquids. In that case, the raw material costs are higher as these reagents are continually required for production. Generally, the raw material cost is lower if enzymatic production methodology is followed or biological reagents like fungi or bacteria are used to treat biomass. In the case of using supercritical substances for production, the cost of the chemicals is included in raw material costs, and the cost of taking the substance to the supercritical stage is included in the operational costs.

Operating cost depends on the production. If the production rate is higher, the operating cost is higher. Apart from the raw material costs, they are costs like electricity or other energy that the plant buys from the grid, labor costs, and packaging costs. If the production is high, the energy also required increases leading to increased requirements from the grid. Some plants build their energy-producing source from renewable sources like solar panels and energy from the same industry's residues, making the plant economical.

Regarding labor cost, if the number of operators increases based on increased production, labor costs are also bound to increase. Conventional biodiesel plants require heating facilities to reach high temperatures. Coal or similar combustible materials are also required based on the production rate. Physical treatment methods of biomass require high temperature, and heavy equipment needs to be operated, making microwave-assisted biodiesel production economically costlier, as the operating cost is higher to produce microwaves.

7.7.3 Fixed cost and maintenance cost

Fixed costs are the costs that occur regardless of the change in the amount of production. These include the salaries of the higher officials, which are essential for the plant to function. Fixed costs also include the cost of the rented land or equipment. The maintenance cost includes transportation and equipment maintenance. Marketing costs are also included in fixed costs if there is a need for marketing the product. Insurance costs of the plant, taxes, and the royalty costs for operation also come under fixed costs. Research and Development (R&D) costs are also included in fixed costs as they require laboratories and other supervisions, which must be paid for regardless of the production rate.

7.7.4 Cost calculation with respect to production rate

Finally, these costs are then integrated, and the cost per unit of production is calculated.

The biodiesel production cost per unit is given by

$$\text{Unit production cost} = \frac{\text{total annual cost}}{\text{annual production amount}}$$

The cost per unit production of biodiesel should be less than the market price of biodiesel for a profitable investment. Generally, the factors, such as interest rate and equipment lifetime, do not contribute much to the unit production cost. These may fluctuate based on the cost of raw materials and the market supply-demand characteristics.

TEA reports analyzed the feasibility, comparison, and cost estimation of various technologies of biodiesel production [69–73].

7.8 Perspectives and conclusion

The recent increase in petroleum prices, the decrease in natural reserves availability, and the environmental concern in the world lead to innovative biofuel production. Today's alarm for shifting the world toward greener fuels and sustainability in the future, the market of biodiesel production will be increased and promising for alternate energy. Edible oils and nonedible oils are contenders for biodiesel feedstocks. Since these nonedible oils do not compete for food or preservation purposes, they have a better future prospectus. Several catalysts and their impact on the kinetics of biodiesel production suggest that the novel technologies, such as microwave-assisted production, ultrasonication, and supercritical fluids, will have increased usage and wide application in the forthcoming years. Several expenses in biodiesel production are reviewed in the TEA, and a generalized approach is followed. As far as the catalysts are concerned, the enzymatic catalysts have great scope in the future, especially when different immobilization methods are being discovered continuously. Among all the renewable energy sources, biomass has an increased chance of widespread application for the agricultural community. Utilizing biofuels promotes reducing carbon footprints as lesser greenhouse gas emissions to save our environment.

References

[1] Math MC, Kumar SP, Chetty SV. Technologies for biodiesel production from used cooking oil—a review. Energy Sustain Dev 2010;14(4):339–45.

[2] Vicente G, Martınez M, Aracil J. Integrated biodiesel production: a comparison of different homogeneous catalysts systems. Bioresour Technol 2004;92(3):297–305.

[3] Lotero E, Liu Y, Lopez DE, Suwannakam K, Bruce DA, Goodwin JG, et al. Synthesis of Biodiesel via Acid Catalysis. Industry & Engineering Chemistry Research 2005;44(14):5353–63.

[4] Van Gerpen J. Biodiesel processing and production. Fuel Process Technol 2005;86(10):1097–107.

[5] Asokan MA, Prabu SS, Prathiba S, Akhil VS, Abishai LD, Surejlal ME. Emission and performance behaviour of flax seed oil biodiesel/diesel blends in DI diesel engine. Mater Today Proc 2021.

[6] Vicente G, Martínez M, Aracil J. A comparative study of vegetable oils for biodiesel production in Spain. Energy Fuels 2006;20(1):394–8.

[7] Atadashi IM, Aroua MK, Aziz AA, Sulaiman NM. The effects of catalysts in biodiesel production: a review. J Ind Eng Chem 2013;19(1):14–26.

[8] Ma F, Hanna MA. Biodiesel production: a review. Bioresour Technol 1999;70(1):1–5.

[9] Zhang J, Chen S, Yang R, Yan Y. Biodiesel production from vegetable oil using heterogenous acid and alkali catalyst. Fuel 2010;89(10):2939–44.

[10] Vishal D, Dubey S, Goyal R, Dwivedi G, Baredar P, Chhabra M. Optimization of alkali-catalyzed transesterification of rubber oil for biodiesel production & its impact on engine performance. Renew Energy 2020;158:167–80.

[11] Urrutia C, Sangaletti-Gerhard N, Cea M, Suazo A, Aliberti A, Navia R. Two step esterification–transesterification process of wet greasy sewage sludge for biodiesel production. Bioresour Technol 2016;200:1044–9.
[12] Sanli H, Canakci M. Effects of different alcohol and catalyst usage on biodiesel production from different vegetable oils. Energy Fuels 2008;22(4):2713–19.
[13] Peiter AS, Lins PV, Meili L, Soletti JI, Carvalho SH, Pimentel WR, et al. Stirring and mixing in ethylic biodiesel production. J King Saud Univ Sci. 2020;32(1):54–9.
[14] Refaat AA. Different techniques for the production of biodiesel from waste vegetable oil. Int J Environ Sci Technol 2010;7(1):183–213.
[15] Xie W, Huang X, Li H. Soybean oil methyl esters preparation using NaX zeolites loaded with KOH as a heterogeneous catalyst. Bioresour Technol 2007;98(4):936–9.
[16] Corma A, Iborra S. Optimization of alkaline earth metal oxide and hydroxide catalysts for base-catalyzed reactions. Adv Catal 2006;49:239–302.
[17] Gawande MB, Pandey RK, Jayaram RV. Role of mixed metal oxides in catalysis science—versatile applications in organic synthesis. Catal Sci Technol 2012;2(6):1113–25.
[18] Niju S, Meera Sheriffa Begum KM, Anantharaman N. Preparation of biodiesel from waste frying oil using a green and renewable solid catalyst derived from egg shell. Environ Prog Sustain Energy 2015;34(1):248–54.
[19] Niju S, Meera Sheriffa Begum KM, Anantharaman N. Clam shell catalyst for continuous production of biodiesel. Int J Green Energy 2016;13(13):1314–19.
[20] Niju S, Meera Sheriffa Begum KM, Anantharaman N. Modification of egg shell and its application in biodiesel production. J Saudi Chem Soc 2014;18(5):702–6.
[21] Niju S, Meera Sheriffa Begum KM, Anantharaman N. Enhancement of biodiesel synthesis over highly active CaO derived from natural white bivalve clam shell. Arab J Chem 2016;9(5):633–9.
[22] Girish N, Niju S, Meera Sheriffa Begum KM, Anantharaman N. Utilization of a cost effective solid catalyst derived from natural white bivalve clam shell for transesterification of waste frying oil. Fuel 2013;111:653–8.
[23] Gupta J, Agarwal M. Preparation and characterizaton of CaO nanoparticle for biodiesel production. In: AIP conference proceedings, Apr 13 2016 (vol. 1724, no. 1). AIP Publishing LLC; 2016, p. 020066.
[24] Venkat Reddy CR, Oshel R, Verkade JG. Room-temperature conversion of soybean oil and poultry fat to biodiesel catalyzed by nanocrystalline calcium oxides. Energy Fuels 2006;20(3):1310–14.
[25] Franssen MC, Steunenberg P, Scott EL, Zuilhof H, Sanders JP. Immobilised enzymes in biorenewables production. Chem Soc Rev 2013;42(15):6491–533.
[26] Yiu HH, Keane MA. Enzyme–magnetic nanoparticle hybrids: new effective catalysts for the production of high value chemicals. J Chem Technol Biotechnol 2012;87(5):583–94.
[27] Leung DY, Guo Y. Transesterification of neat and used frying oil: optimization for biodiesel production. Fuel Process Technol 2006;87(10):883–90.
[28] Patil PD, Deng S. Transesterification of camelina sativa oil using heterogeneous metal oxide catalysts. Energy Fuels 2009;23(9):4619–24.
[29] Adewale P, Dumont MJ, Ngadi M. Recent trends of biodiesel production from animal fat wastes and associated production techniques. Renew Sustain Energy Rev 2015;45:574–88.
[30] Abdullah SH, Hanapi NH, Azid A, Umar R, Juahir H, Khatoon H, et al. A review of biomass-derived heterogeneous catalyst for a sustainable biodiesel production. Renew Sustain Energy Rev 2017;70:1040–51.
[31] Muhammad G, Alam MA, Mofijur M, Jahirul MI, Lv Y, Xiong W, et al. Modern developmental aspects in the field of economical harvesting and biodiesel production from microalgae biomass. Renew Sustain Energy Rev 2021;135:110209.
[32] Tan SX, Lim S, Ong HC, Pang YL. State of the art review on development of ultrasound-assisted catalytic transesterification process for biodiesel production. Fuel 2019;235:886–907.
[33] Ramachandran K, Suganya T, Gandhi NN, Renganathan S. Recent developments for biodiesel production by ultrasonic assist transesterification using different heterogeneous catalyst: a review. Renew Sustain Energy Rev 2013;22:410–18.

[34] Suslick KS, Eddingsaas NC, Flannigan DJ, Hopkins SD, Xu H. The chemical history of a bubble. Acc Chem Res 2018;51(9):2169–78.
[35] Kumar G, Kumar D, Johari R, Singh CP. Enzymatic transesterification of *Jatropha curcas* oil assisted by ultrasonication. Ultrason Sonochem 2011;18(5):923–7.
[36] Adewale P, Dumont MJ, Ngadi M. Enzyme-catalyzed synthesis and kinetics of ultrasonic assisted methanolysis of waste lard for biodiesel production. Chem Eng J 2016;284:158–65.
[37] Tupufia SC, Jeon YJ, Marquis C, Adesina AA, Rogers PL. Enzymatic conversion of coconut oil for biodiesel production. Fuel Process Technol 2013;106:721–6.
[38] Ho WW, Ng HK, Gan S. Advances in ultrasound-assisted transesterification for biodiesel production. Appl Therm Eng 2016;100:553–63.
[39] Suresh T, Sivarajasekar N, Balasubramani K. Enhanced ultrasonic assisted biodiesel production from meat industry waste (pig tallow) using green copper oxide nanocatalyst: comparison of response surface and neural network modelling. Renew Energy 2021;164:897–907.
[40] Veljković VB, Avramović JM, Stamenković OS. Biodiesel production by ultrasound-assisted transesterification: state of the art and the perspectives. Renew Sustain Energy Rev 2012;16(2):1193–209.
[41] Tiwari A, Rajesh VM, Yadav S. Biodiesel production in micro-reactors: a review. Energy Sustain Dev 2018;43:143–61.
[42] Lee CY, Chang CL, Wang YN, Fu LM. Microfluidic mixing: a review. Int J Mol Sci 2011;12 (5):3263–87.
[43] Barredo-Damas S, Alcaina-Miranda MI, Bes-Piá A, Iborra-Clar MI, Iborra-Clar A, Mendoza-Roca JA. Ceramic membrane behavior in textile wastewater ultrafiltration. Desalination 2010;250(2):623–8.
[44] Dubé MA, Tremblay AY, Liu J. Biodiesel production using a membrane reactor. Bioresour Technol 2007;98(3):639–47.
[45] Baroutian S, Aroua MK, Raman AA, Sulaiman NM. A packed bed membrane reactor for production of biodiesel using activated carbon supported catalyst. Bioresour Technol 2011;102(2):1095–102.
[46] Kiwi-Minsker L, Renken A. Microstructured reactors for catalytic reactions. Catal Today 2005;110 (1–2):2–14.
[47] Wen Z, Yu X, Tu ST, Yan J, Dahlquist E. Intensification of biodiesel synthesis using zigzag microchannel reactors. Bioresour Technol 2009;100(12):3054–60.
[48] Tanawannapong Y, Kaewchada A, Jaree A. Biodiesel production from waste cooking oil in a microtube reactor. J Ind Eng Chem 2013;19(1):37–41.
[49] Guan G, Kusakabe K, Moriyama K, Sakurai N. Transesterification of sunflower oil with methanol in a microtube reactor. Ind Eng Chem Res 2009;48(3):1357–63.
[50] Sun J, Ju J, Ji L, Zhang L, Xu N. Synthesis of biodiesel in capillary microreactors. Ind Eng Chem Res 2008;47(5):1398–403.
[51] Lidström P, Tierney J, Watheyb B, Westmana J. Microwave assisted organic synthesis—a review. Tetrahedron 2001;57:9225–83.
[52] Khedri B, Mostafaei M, Safieddin, Ardebili SM. A review on microwave-assisted biodiesel production. Energy Sources A: Recov Util Environ Effects 2019;41(19):2377–95.
[53] Gole VL, Gogate PR. Intensification of synthesis of biodiesel from non-edible oil using sequential combination of microwave and ultrasound. Fuel Process Technol 2013;106:62–9.
[54] Ardebili SM, Hashjin TT, Ghobadian B, Najafi G, Mantegna S, Cravotto G. Optimization of biodiesel synthesis under simultaneous ultrasound-microwave irradiation using response surface methodology (RSM). Green Process Synth 2015;4(4):259–67.
[55] Refaat AA, El Sheltawy ST, Sadek KU. Optimum reaction time, performance and exhaust emissions of biodiesel produced by microwave irradiation. Int J Environ Sci Technol 2008;5(3):315–22.
[56] Amani T, Haghighi M, Rahmanivahid B. Microwave-assisted combustion design of magnetic Mg–Fe spinel for MgO-based nanocatalyst used in biodiesel production: Influence of heating-approach and fuel ratio. J Ind Eng Chem 2019;80:43–52.
[57] Nayak SN, Bhasin CP, Nayak MG. A review on microwave-assisted transesterification processes using various catalytic and non-catalytic systems. Renew Energy 2019;143:1366–87.
[58] Barnard TM, Leadbeater NE, Boucher MB, Stencel LM, Wilhite BA. Continuous-flow preparation of biodiesel using microwave heating. Energy Fuels 2007;21(3):1777–81.

[59] Mueanmas C, Prasertsit K, Tongurai C. Feasibility study of reactive distillation column for transesterification of palm oils. Int J Chem Eng Appl 2010;1(1):77.
[60] Pradana YS, Hidayat A, Prasetya A, Budiman A. Biodiesel production in a reactive distillation column catalyzed by heterogeneous potassium catalyst. Energy Procedia 2017;143:742–7.
[61] Kiss AA, Omota F, Dimian AC, Rothenberg G. The heterogeneous advantage: biodiesel by catalytic reactive distillation. Top Catal 2006;40(1–4):141–50.
[62] Niju S, Meera Sheriffa Begum KM, Anantharaman N. Continuous flow reactive distillation process for biodiesel production using waste egg shells as heterogeneous catalysts. RSC Adv 2014;4(96):54109–14.
[63] Pinto LF, da Silva DI, da Silva FR, Ramos LP, Ndiaye PM, Corazza ML. Phase equilibrium data and thermodynamic modeling of the system (CO2 + biodiesel + methanol) at high pressures. J Chem Thermodyn 2012;44(1):57–65.
[64] Araújo OA, Silva FR, Ramos LP, Lenzi MK, Ndiaye PM, Corazza ML. Phase behaviour measurements for the system (carbon dioxide + biodiesel + ethanol) at high pressures. J Chem Thermodyn 2012;47:412–19.
[65] Hoang D, Bensaid S, Saracco G. Supercritical fluid technology in biodiesel production. Green Process Synth 2013;2(5):407–25.
[66] Lee JH, Kim SB, Kang SW, Song YS, Park C, Han SO, et al. Biodiesel production by a mixture of *Candida rugosa* and *Rhizopus oryzae* lipases using a supercritical carbon dioxide process. Bioresour Technol 2011;102(2):2105–8.
[67] Lozano P, Bernal JM, Vaultier M. Towards continuous sustainable processes for enzymatic synthesis of biodiesel in hydrophobic ionic liquids/supercritical carbon dioxide biphasic systems. Fuel 2011;90 (11):3461–7.
[68] Bernal JM, Lozano P, García-Verdugo E, Burguete MI, Sánchez-Gómez G, López-López G, et al. Supercritical synthesis of biodiesel. Molecules. 2012;17(7):8696–719.
[69] Apostolakou AA, Kookos IK, Marazioti C, Angelopoulos KC. Techno-economic analysis of a biodiesel production process from vegetable oils. Fuel Process Technol 2009;90(7–8):1023–31.
[70] Huang H, Long S, Singh V. Techno-economic analysis of biodiesel and ethanol co-production from lipid-producing sugarcane. Biofuels Bioprod Biorefining 2016;10(3):299–315.
[71] Lee JC, Lee B, Ok YS, Lim H. Preliminary techno-economic analysis of biodiesel production over solid-biochar. Bioresour Technol 2020;306:123086.
[72] Diniz AP, Sargeant R, Millar GJ. Stochastic techno-economic analysis of the production of aviation biofuel from oilseeds. Biotechnol Biofuels 2018;11(1):1–5.
[73] Heo HY, Heo S, Lee JH. Comparative techno-economic analysis of transesterification technologies for microalgal biodiesel production. Ind Eng Chem Res 2019;58(40):18772–9.

Index

Note: Page numbers followed by "*f*" and "*t*" refer to figures and tables, respectively.

C

D

E

F

Printed in the United States
by Baker & Taylor Publisher Services